A Field Guide to

Common South Texas Shrubs

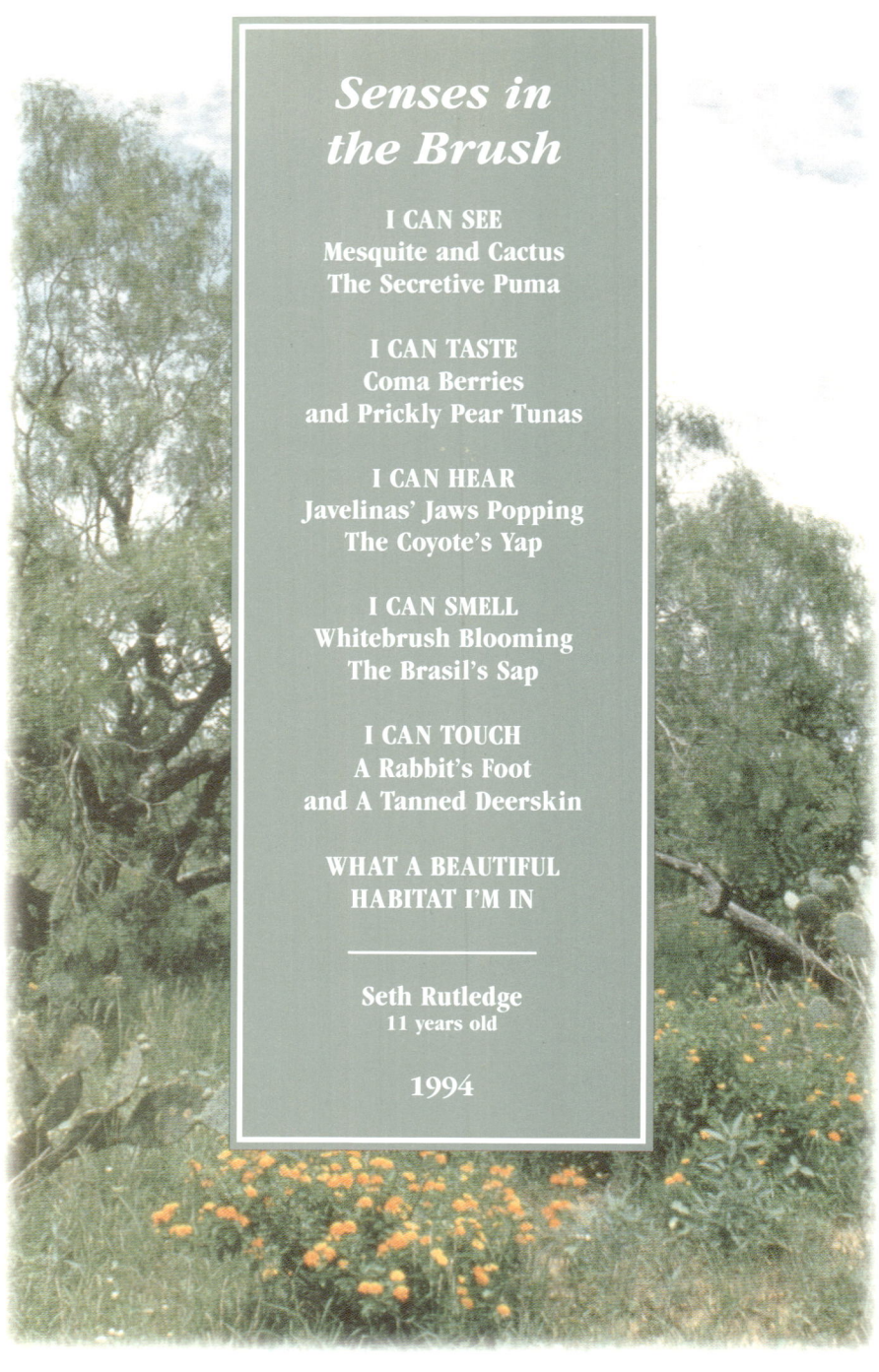

Senses in the Brush

I CAN SEE
Mesquite and Cactus
The Secretive Puma

I CAN TASTE
Coma Berries
and Prickly Pear Tunas

I CAN HEAR
Javelinas' Jaws Popping
The Coyote's Yap

I CAN SMELL
Whitebrush Blooming
The Brasil's Sap

I CAN TOUCH
A Rabbit's Foot
and A Tanned Deerskin

**WHAT A BEAUTIFUL
HABITAT I'M IN**

Seth Rutledge
11 years old

1994

A Field Guide to
Common
South Texas
Shrubs

Richard B. Taylor
Jimmy Rutledge
Joe G. Herrera

Project Coordinators:
Publisher: Georg Zappler
Art Director: Pris Martin
Design: Suzanne Davis
Cover Design: Suzanne Davis
Illustrations: Rob Fleming
Printing: Mike Diver

Funding for
A Field Guide to Common South Texas Shrubs
was partially provided by Pittman-Robertson W-129-M,
Federal Aid in Wildlife Restoration

Another "Learn about Texas" publication from
TEXAS PARKS AND WILDLIFE PRESS
ISBN: 1-885-696-14-0

Contents

This book is dedicated to
Burns & Adeline Taylor;
Marcy, Seth, Luke & Jess Rutledge;
and Walter Elling

Acknowledgements

The authors are extremely grateful to:
Dr. Tim Fulbright
Texas A&M University, Kingsville,
and
Dr. Chester M. Rowell, Jr.
Adjunct Professor of Biology,
Sul Ross State University,
for their time and expertise
in reviewing and
editing this manuscript.

We wish to thank:
Dr. Eric Hellgren
Caesar Kleberg Wildlife Research
Institute, Texas A&M University,
Kingsville, for his assistance in
the nutritional analysis of
some of the plants,
and
Dr. Grady Webster
Professor Emeritus of Botany,
Curator – Tucker Herbarium,
University of California, Davis,
for reviewing the manuscript.

We extend a special thanks to
Texas Parks and Wildlife's
Don B. Frels, Regional Director;
W. J. Williams, District Supervisor;
and *David Synatszke*, Area Manager
of the Chaparral Wildlife Management
Area, for their faith, encouragement, and
especially patience in this project.

We would like to thank
*Steve Nelle, Patty Leslie, Jim Hillje,
Lee Miller and Randy Fugate*
for their field and research assistance.
And finally, to the landowners and
sportsmen and women who provided the
natural resources, desire, and reasons
to publish this field guide, we extend
further thanks.

Preface

This plant-identification field guide is an attempt by the authors to assist sportsmen and women, landowners, managers, biologists, ecologists and anyone else interested in the identification of woody vegetation in south Texas. Our objective is to produce an informative, easy-to-use plant-identification field guide, providing the most current knowledge and nutritional information available on selected woody plants and cacti important to wildlife in south Texas. As range and wildlife students at Texas A&I University and later as Texas Parks and Wildlife Department wildlife biologists, it was apparent to us that a field guide to the common woody plants of south Texas was needed. More comprehensive plant-identification books currently available include Robert E. Vines' *Trees, Shrubs, and Woody Vines of the Southwest* and D.S. Correll and M.C. Johnston's *Manual of the Vascular Plants of Texas*. While truly outstanding and invaluable as reference sources, collectively they contain almost 3,000 pages and weigh about 10 pounds, making them impractical for field use.

Over 281 species of woody plants and 32 species of cacti are recognized in the south Texas ecological region. The vast majority of these are found in the lower Rio Grande Valley, which is part of the subtropical Tamaulipan biotic province. Many of the plant species in this area reach their northernmost boundary here. The 44 plants described in this guide represent an estimated 75% of the overall brush and cacti biomass of the south Texas ecological region, excluding the lower Rio Grande Valley.

The scientific names used in this book were cited from *A Synonymized Checklist of the Vascular Flora of the United States, Canada and Greenland* by John T. Kartesz, 1994. We grouped the plants into thorned and thornless categories and alphabetized them by family. Distinguishing characteristics have been italicized for easy reference. Similar species are also noted to further assist in identification. In this guide, plants are not ranked by importance, because their value to wildlife can differ from ranch to ranch, depending on availability, location, soil type and land-management practices. For plants not found in this guide and for more information, a list of additional references is included. In addition, local Natural Resources Conservation Service staff, the County Agricultural Extension Agent, and Texas Parks and Wildlife biologists can be of assistance.

Richard B. Taylor

Jimmy Rutledge

Introduction

The South Texas Plains constitute a triangular region, roughly south and east of a line from Del Rio to San Antonio to Rockport. Also called the Rio Grande Plains or south Texas "brush country," the region encompasses about 20.5 million acres, covering fifteen counties in their entirety and portions of fourteen others. Major ranches such as King, Kenedy, Callaghan, Piloncillo, San Pedro, Briscoe, Chittum and Farias are located in the area.

The topography varies from generally level to a gently undulating plain that drains into the Gulf of Mexico by way of the Rio Grande and Nueces rivers and their tributaries. Elevation ranges from sea level along the coast to nearly 1,000 feet in the northwestern portion. Soils are very diverse and range from clays to fine sands and from calcareous to slightly acidic. Since specific soils generally support distinct plant communities, common vegetation communities can generally be characterized on the basis of soil maps. Frequently, certain plant species can be used to identify and describe habitat quality and the wildlife values of a site.

The climate of the area is generally mild with average annual temperatures ranging from 66 to 74 degrees Fahrenheit. Summer temperatures often exceed 100 degrees and severe frosts occasionally occur during the winter. The average growing season lasts 340 to 360 days. Prevailing winds are from the southeast, bringing warm moist air from the Gulf of Mexico. There is a pronounced rainfall gradient across the region, with average rainfall about 33" on the eastern edge and 17" on the western edge. Peak rainfall occurs in May, with a secondary peak in September. Droughts are common and "normal" rainfall is frequently significantly above or below the historical average. Extreme variability in rainfall and hot temperatures are important factors that influence habitats of south Texas.

Spanish explorers who traveled across this region left documentation describing the landscape and vegetation. Contrary to popular belief, the entire area was not a continuous prairie of "stirrup high" grasses. Although grasslands apparently dominated the landscape, woody plants (trees and shrubs) were often present in thickets, upland areas, major drainages and river bottoms. Natural fires helped to maintain the region as a grassland or savannah, and reduced woody plant densities.

Early settlers of the region depended on livestock to make their living. The invention and introduction of barbed wire allowed settlers to fence livestock. Fencing and unrealistic expectations of grazing capacity led to overgrazing. Although the area is historically known for its thriving cattle industry, from the late 1860s to the 1890s sheep played an extremely important role in shaping present-day south Texas. Corpus Christi was the nation's largest export center for wool in 1880. In 1889, four of the top sheep-producing counties in Texas were located within the South Texas Plains.

The sheep industry declined around the turn of the century and cattle have since dominated the livestock industry.

Brush densities increased during the late nineteenth and early twentieth centuries due to a variety of factors, which probably included overgrazing, lack of natural fires, soil compaction and periodic droughts. Originally, increased brush density was regarded as detrimental by ranchers, and attempts to control it were intensive and widespread. The earliest attempts at brush control and range reseeding began during the late 1930s and early 1940s, with cabling and chaining. As ranchers generated income from livestock production, they also practiced rootplowing, roller chopping, Rome discing and chemical spraying to control brush and to increase grass production. Research into the effects of brush management by wildlife biologists in the late 1960s indicated that extensive brush control was detrimental to wildlife. White-tailed deer hunting increased in popularity during the late 1950s, 1960s and 1970s, giving landowners an economic incentive to provide quality habitat for white-tailed deer. Consequently, brush removal has been reduced or applied in a manner that is not detrimental to wildlife, thus improving the habitat for deer and other wildlife species. Research has shown deer prefer forbs; however, woody plants and cacti become an important staple in drought-prone western south Texas. Identification of key food plants then becomes an important aspect of evaluating habitat, range condition and ecosystem health. Additionally, knowing the nutritional value of plants can also help in range analysis.

Protein in an animal's diet is essential for growth, maintenance and reproduction. Crude protein (CP) is an estimate of a plant's protein content expressed as a percentage of the total plant, but is not what is actually available to the animal. Certain plants (e.g. *Acacias* spp.) have defenses against consumption which include structural deterrents such as thorns and/or chemical weapons. The chemical defenses are secondary compounds which sometimes interfere with digestibility and/or are poisonous to the animal. Digestible protein (DP) is the amount of protein in the plant that is actually digested and utilized by the animal. Digestible dry matter (DMD) is the percentage of ingested food actually absorbed into the animal's system. This percentage includes proteins, along with carbohydrates. These values are found in the nutritional chart provided at the end of this book.

Brush management can affect the nutritional value and chemical composition of plants. Proper brush management can benefit wildlife by improving the nutritional value of plants or by just creating desired habitat, sometimes both. There are many techniques used for manipulating or managing brush densities and diversity, including mechanical and chemical means, as well as fire. Fire or controlled burning is an excellent technique for brush management and extremely beneficial to wildlife. The most limiting factor in south Texas is adequate herbaceous ground cover to fuel the fire. Controlled burning is often used as a follow-up treatment after mechanical or chemical manipulation.

Brush-management patterns such as strips, blocks, zigzags, contours and mosaics are also important. Elaborate patterns are more expensive but are generally more beneficial to wildlife, due to the creation of greater landscape diversity and increased "edge effect." A minimum of 250 yards on either side of creeks and drainages should be left intact to provide travel corridors for wildlife. Brush removal should not exceed 25– 30 percent of the overall area. Soil types, topography, vegetation types, climate, clearing patterns, amount of removal and pre-management and post-management usage must all be considered when managing brush. A certain density of brush may provide good habitat for some wildlife species while not for others.

Livestock-management issues, including stocking rates, grazing systems, water availability and supplemental feeding distribution, also affect wildlife habitat. Livestock grazing has a greater effect on wildlife habitat than any other factor in terms of acres impacted. Livestock utilize brush for food and shelter to varying degrees, depending on season and forage availability. Proper stocking rates and grazing systems benefit the producer and the resource. A multi-herd, multi-pasture rotational system ranked high for simultaneous production of livestock, white-tailed deer, bobwhite quail and turkeys in one scientific study. Lower stocking rates increase ground cover for ground-nesting birds and mammals, by decreasing potential nest destruction and predation. Additionally, adequate ground cover lowers ground temperatures and increases humidity, providing for better productivity of ground-nesting vertebrates such as quail, turkeys and tortoises.

"Although brush may be brush to the casual observer, woody plants vary greatly in quality and value to both game and livestock."
—Val Lehmann
Forgotten Legions: Sheep in the Rio Grande Plains of Texas

Ecoregions of South Texas

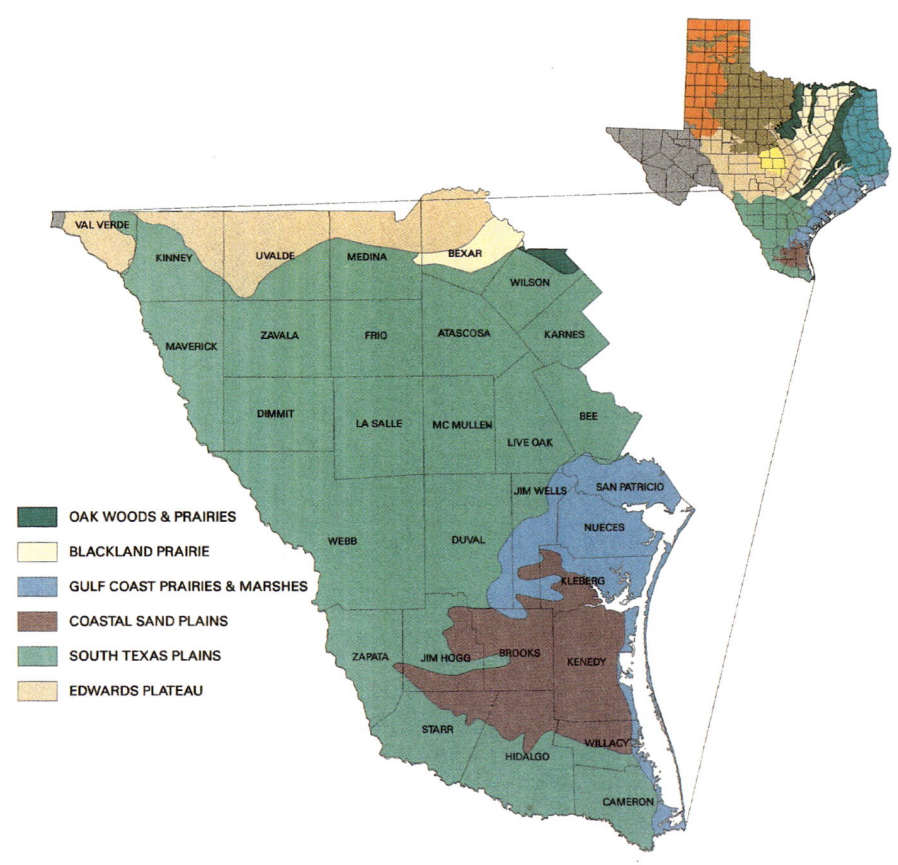

OAK WOODS & PRAIRIES
BLACKLAND PRAIRIE
GULF COAST PRAIRIES & MARSHES
COASTAL SAND PLAINS
SOUTH TEXAS PLAINS
EDWARDS PLATEAU

Information based upon the Natural Heritage Policy Research Project, 1978.
Map graphic produced by Cynthia Banks, Texas Parks and Wildlife, GIS Lab.

Thorned Plants

Cactus Family   CACTACEAE

Prickly Pear
Opuntia engelmannii

(nopal)

Description

An erect or spreading, thick-padded, thicket-forming cactus (3'–10') with a cylindrical trunk. The yellow, orange or red flowers are produced from April to June. Red-to-dark-purple fruit, also called tunas or pear apples ($1/2$"–$2^1/2$"), ripen from July through September.

Prickly pear is the most common cactus found throughout south Texas in all soil types and in most plant communities.

*Crude Protein Value				
	spring	summer	fall	winter
pads:	2%-13%	6%-7%	7%-10%	2%-6%
fruit:	N/A	6%-8%	N/A	N/A

Range in value results from variation among studies influenced by climate, soil types, plant growth stage, etc.

Values

Prickly pear has a high moisture content (80%) and is extremely valuable to wildlife for food, water and cover. White-tailed deer, javelina and feral hogs eat the pads and fruit. The fruit is relished by many mammals, birds and reptiles, including rodents, rabbits, ground squirrels, coyotes, raccoons, bobwhite quail, scaled quail, mourning doves, white-winged doves, turkey, curve-billed thrashers, golden-fronted woodpeckers, Chihuahuan ravens, wrens and the Texas tortoise. Quail, rodents, skunks and many reptiles utilize prickly pear for cover or nesting. Cactus wrens, roadrunners, thrashers and such rodents as the wood rat frequently nest in prickly pear. Wood-rat middens are common in and around the base of the plant.

Cattle and goats occasionally feed on young prickly pear pads and the fruit, and also eat the mature pads. During droughts ranchers burn the spines off the cactus to supplement cattle diets.

The fruit and pads of prickly pear are edible by humans also. The fruit can be made into jelly, wine and syrup. Young pads, called *nopalitos*, can be boiled and then fried with eggs to make a breakfast meal. Historically, American Indians made tea from the fruit to cure gallstone ailments, and made poultices from the pads to relieve pain from cuts, abrasions, swollen bruises and toothaches. It is also reported that the pads were sometimes dried and stitched together to make bags for food and water. The juice of the pads, when boiled and mixed with tallow, was used in the hardening of candles. Prickly pear is easily established and can be successfully used in erosion control.

Tasajillo
Opuntia leptocaulis

(Christmas cactus, rat-tail cactus,
pencil cactus, turkey pear)

Description

A thicket-forming, cylindric or pencil-stemmed cactus that usually grows erect and sometimes among other plants. The slender branches are jointed and easily detached. The small, greenish-yellow flowers ($^1/_2$"–$^3/_4$") are produced on stems from May to August, and *open only late in the evening*. The numerous

Crude Protein Value				
	spring	summer	fall	winter
stems:	8%	8%	8%	8%
fruit:	N/A	8%	N/A	N/A

small, bright red or orange oblong fruits give the cactus a "Christmas" look. Tasajillo is a very common cactus, and is widely scattered throughout Texas. The plant is found in all soil types but prefers sandy and heavier bottomland soils. It is frequently found along fence rows and under trees where birds have deposited the seeds.

Values

Tasajillo is a good wildlife food plant. White-tailed deer eat the fruit and young succulent growth. Most birds, including bobwhite quail and wild turkey as well as small mammals, eat the berries. Cactus wrens nest in the larger plants.

Although tasajillo is a good wildlife plant, it is of little value for livestock or humans.

Allthorn
Koeberlinia spinosa

(crucifixion thorn, junco)

Description

A thorny, multiple-branched shrub forming a mostly stiff, leafless, *green, tangled mass of spines*. Allthorn is leafless throughout most of the year except following a rain, when small, alternate leaves appear for a short time. The tiny, greenish-white flowers are four-petaled and the clustered fruit consists of 4–8 small black berries.

Allthorn is a minor component of mixed-brush chaparral communities. It is found on most soil types, from gravelly soils or caliches, to clay and sandy arroyos.

Values

New regrowth is sweet to taste and is browsed by small and large mammals. Birds, such as the scaled and bobwhite quail, and small mammals, such as the jackrabbit, eat the fruit. Allthorn also provides protective cover for many small mammals and reptiles. It can cause minor injuries, such as skin tears or punctures, to livestock.

7

Blackbrush
Acacia rigidula

(chaparro prieto)

Description

A medium-sized, deciduous, thorny shrub (3'–15') with multiple basal stems and whitish-to-dark-gray branches. It frequently forms dense thickets. The twice-compounded leaves have 2–5 pairs of heavily ribbed leaflets on 1–2 pairs of pinnae. Fragrant white or light yellow flowers are produced from February to April in clustered, oblong spikelets. The linear, flattened legume (2"–3^1/$_2$" L X ^{1}/$_4$" W) is constricted between the reddish-brown-to-black seeds. The light gray, straight spines (^{1}/$_2$"–1^1/$_2$") are paired at the nodes.

Blackbrush is a common component of south Texas brushland, and is frequently found on sandy, gravelly or limestone caliche ridges and hills; it is

often associated with guajillo. The shrub is also found in a variety of soils and mixed-brush communities.

Values

White-tailed deer browse the leaves, eat the legumes and relish the flowers. The seeds are eaten by a variety of birds, including bobwhite quail. Many birds and small mammals may be found around blackbrush, utilizing it for cover and protection. Certain birds, such as cactus wrens and scissor-tailed flycatchers, nest in it. Hummingbirds and bees frequent the flowers, making blackbrush a source of honey.

It is a relatively poor browse plant for livestock and the spines may cause minor injury to mouth parts.

Blackbrush is drought tolerant and can be used as a landscaping plant in dry areas, xeriscapes and rock gardens.

Legume Family   FABACEAE

Catclaw
Acacia greggii

(Texas mimosa, uña de gato, wait-a-while)

Description

A thorny, colony-forming, deciduous mid-sized shrub (3'-10') with numerous slender branches having recurved thorns ($1/4$") *resembling catclaws* that make the plant almost impenetrable. The tiny, twice-compounded leaves have 3-7 pairs of leaflets on 1-3 pairs of pinnae. The creamy yellow flowers ($1^1/4$"-$2^1/2$") are produced from April through October in oblong spikelets. The thin, flat legume (2"-$5^1/2$" L X $1/2$"-$3/4$" W) often appears straight, curved or twisted with small, flat, brownish seeds. A similar species is Wright's acacia, which is distinguished by its growth form and differences in the legume.

Catclaw is a common component of south Texas brushlands, and is found in a variety of habitats and soil types, from dry arroyos and valleys to sandy or gravelly hills and slopes. It is often associated with other acacia species and a variety of south Texas shrubs.

Values

Tender new growth provides browse for white-tailed deer and rabbits, especially when other food is limited. The seeds are eaten by a variety of small mammals and birds, including scaled and bobwhite quail. Catclaw provides protective cover for wildlife, including quail and small mammals, and serves as a nesting and loafing tree for a number of bird species. It is a food plant for butterfly larvae and a nectar source for adult butterflies. Several insects, living on this plant, produce a resinous substance that is used in making varnish and shellac.

*Crude Protein Value				
	spring	summer	fall	winter
leaves:	19%–30%	16%–19%	16%–19%	17%

*Range in value results from variation among studies influenced by climate, soil types, plant growth stage, etc.

Cattle browse on young growth, especially when other forage is limited.

American Indians ground the legumes to make a mush or cake to ease back pain. The wood has been used for fuel, and bees produce honey from the flowers. Catclaw is drought tolerant and can be used as an ornamental or landscape plant in dry areas, xeriscapes and rock gardens. Due to its thorny, impenetrable structure, it also makes a good hedge plant.

Guajillo
Acacia berlandieri

(Berlandier acacia, thornless catclaw)

Description

A small-to-medium-sized shrub (4'–10') with multiple basal stems flaring outward to form a rounded crown. Guajillo frequently forms dense thickets and has small, inconspicuous, slightly-recurved-to-straight thorns. The *narrow, delicate and fernlike*, twice-compounded leaves (4"–6") have 30–50 pairs of leaflets on 5–12 pairs of pinnae. The small, round ($1/2$"+), creamy-white-to-yellow, fragrant flowers are produced from November through April. The legumes (4"–6") mature in June and July, with 5–10 broad, dark brown seeds. The legumes shatter, spreading beans quickly; however, the previous year's fruit may remain on the plant.

Guajillo is a common component of south Texas brushlands, and is found in a variety of habitats and soil types, most frequently in shallow sandy, gravelly or limestone caliche hills or ridges. It is often associated with blackbrush, tasajillo, cenizo and prickly pear.

*Crude Protein Value				
	spring	summer	fall	winter
leaves:	27%–28%	20%–22%	19%–22%	17%–21%
beans:	17%	N/A	N/A	N/A

Range in value results from variation among studies influenced by climate, soil types, plant growth stage, etc.

Values

Guajillo is browsed by white-tailed deer. The seeds are consumed by feral hogs and scaled quail. Small mammals utilize the guajillo thickets for cover and protection.

Guajillo is browsed by cattle, sheep and goats but is low in digestibility. It can become toxic for sheep and goats if excessive amounts are consumed over a long period of time.

Guajillo nectar makes an excellent honey and the wood has been used for fuel, tool handles and small wooden articles. Gum and dark gel have been extracted from this shrub. It is drought tolerant and can be used as an ornamental, landscape or hedge plant in dry areas, xeriscapes and rock gardens.

Honey Mesquite
Prosopis glandulosa

(mesquite)

Description

A spiny, deciduous shrub or small tree (10'–30') with crooked, grayish branches and drooping foliage that forms a roundish crown. The smooth, green, alternate leaves are twice compounded, with 6–20 leaflets on 1–2 pairs of pinnae usually emerging after the last cold front. The yellowish green, cylindrical flowers (2"–3") are produced from April through September. The oblong legumes (4"–10") are arranged in loose clusters that ripen into a shiny light brown in the late summer. The trunk of older trees has rough, deep-cracked, sap-stained bark, a twisted appearance and reddish-brown heartwood.

Mesquite is the most common tree found throughout south Texas in all plant communities and soil types. It is an aggressive invader and frequently is the first plant to reestablish following clearing or disturbance. It often forms dense thickets that may restrict some types of land use.

*Crude Protein Value				
	spring	summer	fall	winter
leaves:	23%–32%	16%–17%	26%	21%
beans:	11%–13%	N/A	N/A	N/A

Range in value results from variation among studies influenced by climate, soil types, plant growth stage, etc.

Values

Mesquite is extremely important to wildlife in south Texas. It provides browse for white-tailed deer and the beans are eaten by many mammals, including white-tailed deer, javelina, feral hogs, coyotes, ground squirrels and other rodents. The seeds are eaten by mammals, including rock squirrels, ground squirrels, coyotes, skunks, rodents, jackrabbits and white-tailed deer, and by birds, including bobwhite quail, scaled quail, white-winged doves and mourning doves. The trees provide nesting, roosting and loafing cover for a variety of birds, including mourning doves, white-winged doves, scissor-tailed flycatchers and chachalacas. Mesquite fixes nitrogen into the soil, which benefits other plant species growing under it; it also often provides a cooler microclimate under its canopy, which benefits wildlife, especially during the summer. The flowers provide good bee food and honey. Mesquite is a good food plant for butterfly larvae and a nectar source for adult butterflies.

Livestock browse on the foliage and legumes. When they pass through their digestive tracts, the seeds may establish new plants where they are deposited, generally away from under the canopies of mature mesquite trees. Excessive consumption of the beans may be toxic.

Mesquite wood is widely used for a variety of purposes, including as firewood, furniture, posts and flooring. Mesquite is drought, disease and insect tolerant, making it a good ornamental and landscape tree. It also provides good urban wildlife habitat. Historically, the legumes were an important source of food for American Indians, who ground them into a flour to make bread. By fermenting them, they also made an intoxicating beverage. Mesquite also produces a black dye and a cement for mending pottery. Medicinally, gum from the bark was eaten like candy or dissolved in water to treat dysentery, sore throats and open wounds.

Legume Family   FABACEAE

Huisache
*Acacia minuta**

(sweet acacia, honey ball, uña de cabra)

Description

A fast-growing, short-lived (30–50 years), deciduous, thorny shrub or small tree (10'–30') with multiple basal stems flaring upward to form a spreading or flattened crown. The twice-compounded leaves (1"–4") have 10–25 pairs of leaflets on 2–8 pairs of pinnae. The round, yellow-gold, fragrant flowers are produced from February to March, and give it a densely flowered, showy appearance. The reddish brown, purple or black legumes (2"–3" L X $^1/_2$"–$^2/_3$" W) are thick, oblong, straight or curved with green-to-brown seeds in solitary compartments. The branches have paired spines at the leaf base. Huisache is often confused with *twisted acacia*, but can be distinguished by its growth form and differences in the legume.

Huisache is a common component of the south Texas brushland. Found in a variety of soil and habitat types, it occurs more frequently in deep, poorly drained sandy or clay lowland areas. Huisache is an aggressive invader, especially after brush clearing or other types of disturbance have occurred.

*Referred to as *Acacia farnesiana* or *Acacia smallii* in some sources.

16

Crude Protein Value				
	spring	summer	fall	winter
leaves:	N/A	23%	N/A	N/A
beans:	N/A	18%	N/A	N/A

Values

Huisache is browsed by white-tailed deer, which also, together with javelina, eat the fruit. Many birds, including scaled and bobwhite quail, feed on the seeds. It is commonly used for nesting, loafing and cover by many birds, including mourning doves. Rodents sometimes gnaw the stems near watering areas. The flowers provide nectar which attract bees and butterflies.

Huisache provides limited forage for livestock, especially in winter.

Huisache is drought and water tolerant and can be used as an ornamental and landscape plant. The flowers provide a desirable honey and have also been made into an ointment for headaches and into tea for indigestion. Various parts have also been used medicinally to treat dysentery and skin disease. The wood is used for many things, including firewood, posts and other wooden-ware products, and is used in various other processes including tanning, dying and ink and glue making. It also provides oil for perfumes.

Palo Verde
Parkinsonia texana

(border palo verde)

Description

A smooth, *green-barked*, deciduous, very thorny shrub or small tree (4'-10'). The small (³/₄"-1"), light-bluish-green, twice-compounded leaves have 2-3 paired, oblong leaflets, often dropping their leaves in the summer and releafing after rains. *At first glance, it often appears semi-leafless.* The five-petaled, ruffled flowers (1" long) are yellow with a red spot, and the legume (1"-2" L X ¹/₄"-¹/₂" W) is dark brown and flattened. The spines are short and straight at the nodes. During droughts, the green bark allows it to carry on photosynthesis until it releafs. Similar in appearance to *retama*, except the leaves are not found on a long rachis.

Crude Protein Value

	spring	summer	fall	winter
leaves:	24%	N/A	N/A	N/A

Palo verde occurs in moderate densities in south Texas, and is found in sandy loams, clay soils and shallow, well-drained rocky soils. It frequently forms loose colonies.

Values

White-tailed deer, jackrabbits and other small mammals browse palo verde. The seeds are utilized by many mammals, including deer, javelina, feral hogs and such rodents as the kangaroo rat, and also by birds. Hummingbirds, butterflies and bees often seek the flowers.

Cattle eat the legumes. Palo verde has some ornamental and landscape value, especially on hot, dry, poor soils, and provides color in winter. The legumes can be made into a palatable flour and bees make honey from the flowers. The wood can be used for fuel.

Retama
Parkinsonia aculeata

··

(Jerusalem-thorn, horsebean, crown of thorns, lluvia de oro)

Description

A deciduous, smooth, green-barked, slender-branched, thorny shrub or small tree with *feathery foliage* and a drooping, rounded crown (10'-15'+). The leaves are twice compounded on a *long, flat rachis (8"-16")* with many leaflets. The fragrant, five-petaled yellow flowers have a red or orange center or spot on one petal and are produced in drooping, showy clusters throughout the summer, especially after rains. The linear, brown-to-orange or reddish legume ($1^1/2"$-4") is constricted and flattened, and contains 1-8 small seeds. During droughts, the green bark allows it to carry on photosynthesis until it releafs. Retama has very sharp recurved spines and is similar to *palo verde*.

Retama is commonly found throughout south Texas in *moist, poorly drained* or disturbed sandy or limestone

Crude Protein Value				
	spring	summer	fall	winter
leaves:	20%	N/A	N/A	N/A

soils but can be found in a variety of sites and is frequently associated with mesquite-granjeno communities.

Values

White-tailed deer browse the foliage, especially new regrowth after a burn or disturbance, and, along with other mammals, eat the legumes. The seeds are eaten by some birds, including bobwhite quail. Bees are attracted to the flowers.

The leaves and legumes are eaten by cattle. Livestock browse new regrowth, as well as the branches during droughts.

Retama can be used for fuel and has been used in paper making. Historically, American Indians made a coarse flour from the seeds for food, and in Mexico the branches and leaves were made into a medicinal tea to treat diabetes and fever. Retama can also be used as a hedge plant and as an ornamental or landscape tree that provides showy color when flowering and that requires little maintenance.

Texas Ebony
Pithecellobium ebano

(ebony apes earring, ebano)

Description

A densely foliaged, thorny evergreen shrub or tree (15'–30') with zigzagged branching and a very dense, dark canopy. The *thick, dark green*, alternate, twice-compounded leaves have 3–6 pairs of leaflets on 2–3 pairs of pinnae. The cream or yellow, fragrant, long, cylindrical flower clusters are usually produced from June to August. The dark-brown-to-black, straight or slightly curved, thick, flat, *woody, hard* *legumes (4"–6" L X 1"–1¹/₂" W) have bean-shaped, reddish-brown seeds.*

Texas ebony is frequent in the chaparral brush of the coastal part of the Rio Grande Plains, becoming infrequent north to just south of Laredo. It is associated with several plant communities, such as mesquite-blackbrush and mesquite-granjeno.

Crude Protein Value				
	spring	summer	fall	winter
leaves:	23%	20%	23%	22%
beans:	22%	N/A	N/A	N/A

Values

White-tailed deer browse on the foliage. The seeds are eaten by white-tailed deer, javelina, feral hogs, rodents and other small mammals. Texas ebony provides an excellent nesting, loafing and roosting tree for birds, including mourning doves, white-winged doves and songbirds. Bees are attracted to the flowers.

Texas ebony is valuable as shade for livestock.

The wood is extremely valuable and is used for posts, fuel, cabinets, small furniture, and for art objects such as wood carvings. It is an excellent ornamental, landscaping, hedge and shade tree, as well as an urban or backyard wildlife-habitat tree. The seeds are made into jewelry. In Mexico, the seeds are eaten when green and are also roasted when ripe and used as a coffee substitute.

Legume Family   FABACEAE

Twisted Acacia
Acacia schaffneri

(huisachillo, Schaffner acacia)

Description

A thorny, relatively short-lived, low-spreading deciduous shrub (4'–12') usually with many purplish-to-brown, bent and twisted stems that flare outward from the base and form a rounded crown. The leaves are twice compounded, with 10–15 pairs of leaflets on 2–5 pairs of pinnae. The yellow-to-orange, fragrant flowers are in round clusters ($1/4$"–$3/8$"), and the *narrow, twisted* legumes ($2^1/2$"–5") are black, velvety and slightly compressed. A similar species is *huisache*, which is distinguished by its growth form and shorter, thicker legumes.

*Crude Protein Value				
	spring	summer	fall	winter
leaves:	17%–22%	18%–20%	20%–22%	16%–17%
beans:	10%	N/A	N/A	N/A

Range in value results from variation among studies influenced by climate, soil types, plant growth stage, etc.

Twisted acacia is a common component of south Texas brushland and is found in a variety of soils and habitat types in association with most mixed-brush communities. It is among the first shrubs to reestablish in root-plowed or disturbed areas.

Values

Twisted acacia is browsed by white-tailed deer. Feral hogs and javelina eat the beans, and certain birds, including quail, eat the seeds. Many birds and small mammals use it for loafing and as a protective cover, and birds nest in the older, larger plants.

Sheep and goats occasionally browse the foliage.

Twisted acacia can be used as landscape, ornamental or hedge plant in xeriscapes and rock gardens.

Brasil
Condalia hookeri

(bluewood condalia, capul negro, capulin)

Description

A spiny, evergreen shrub or small tree (6'–15'), frequently multiple-trunked, having an irregularly shaped crown. The small (<1"), *shiny, light lime-green leaves are alternate on pale-gray-to-brownish branches that end in sharp spines*. It has small, inconspicuous, greenish flowers and small, purple-to-black fruit that ripens throughout the summer.

Brasil is a major component of south Texas brush and is commonly found in various mixed-brush habitat types and drier soils.

Values

Brasil is a valuable wildlife food plant because the fruit ripens throughout the season. It is eaten by many mammals,

	spring	summer	fall	winter
leaves:	21%–24%	13%–17%	17%–18%	16%–18%
fruit:	8%	N/A	N/A	N/A

*Range in value results from variation among studies influenced by climate, soil types, plant growth stage, etc.

including coyotes, squirrels, raccoons, gray fox and opossums, and by most game and non-game birds, including bobwhite quail, scaled quail, white-winged and mourning doves, orioles, cardinals, thrashers and woodpeckers. The larger trees provide nesting, roosting and loafing areas for birds, and the thickets provide protective cover for most mammals and birds.

The shrub has low browse value for cattle.

The wood has been used for fuel and to make a light red, pink or blue dye. The flower pollen serves as bee food, and jelly and wine is made from the fruit. Brasil is commonly used as a landscape, ornamental and hedge plant, and may also be used to improve backyard habitats for wildlife.

Knife-leaf Condalia
Condalia spathulata

(squawbush, costilla)

Description

A very *spiny, impenetrable, low and wide, irregularly dome-shaped, clump-forming, evergreen shrub (10'-20' diameter clump) with grayish-green branches.* Knife-leaf condalia has small, narrow, alternate leaves ($1/4$"–$1/2$"), inconspicuous, greenish flowers and a round, black, edible fruit. It is similar in appearance to *lotebush* and *amargosa*, except for the clump growth form.

Knife-leaf condalia is a minor component in mixed-brush communities on dry gravelly or caliche hillsides, in shallow soils and along arroyos in dry, open, brushy areas. The shrub is frequently associated with mesquite and prickly pear.

Values

New growth is occasionally browsed by white-tailed deer and livestock. The fruit is eaten by certain birds, including quail, and thickets provide excellent protective cover to many small mammals and birds.

Lotebush
Ziziphus obtusifolia

. . . .

(gumdrop tree, Texas buckthorn, clepe)

Description

A deciduous, spiny, multiple-branched stiff shrub (3'–6') with *grayish-green, thorn-tipped branches* and tiny, inconspicuous, green, five-petaled flowers. The shiny, green, alternate leaves ($1/2$"–$1^1/2$") are linear or narrow-oblong-shaped and occur on the spines. The fruit is small, round ($1/3$"–$1/2$"), black and solitary. It is similar to *amargosa* and *knife-leaf condalia*.

   Lotebush is a seldom-abundant but common component of shrub communities throughout south Texas and occurs in a variety of soil types and mixed-brush communities. The shrub prefers drier habitats, and common associations include brasil, mesquite and prickly pear.

Values

The leaves are occasionally browsed by white-tailed deer and the fruit is eaten by many birds and mammals, such as the gray fox, raccoon, coyote and chachalaca. Lotebush serves as cover and protection for many rodents and quail, and is also used for loafing. Birds such as the cactus wren occasionally nest in lotebush.

30

	spring	summer	fall	winter
leaves:	18%–24%	15%–19%	16%–20%	12%–15%

*Range in value results from variation among studies influenced by climate, soil types, plant growth stage, etc.

Cattle, sheep and goats browse on the foliage; however, the spines may cause injury to the mouth.

The root serves as a soap substitute and as a treatment for wounds of domestic animals.

Citrus Family   RUTACEAE

Lime Prickly-Ash
Zanthoxylum fagara

(colima, uña de gato)

Description

A prickly, aromatic, intricately branched evergreen
shrub (5'–20') with small, greenish-yellow flowers
that are produced from March to June, and that
generally appear after the leaves. The bright green,
oblong leaves are alternate, odd compounded and
on a *broad winged rachis, and are aromatic and
bitter to taste*. The small, reddish-brown-to-black,
roundish, single-seeded fruit ($1/8"–1/4"$) is smooth
and shiny, and ripens in late summer or early fall.
The thorns are recurved; a similar species is
therefore *catclaw*. *Littleleaf sumac* has a winged
rachis, but is thornless.

32

*Crude Protein Value				
	spring	summer	fall	winter
leaves:	17%–23%	12%–18%	6%–19%	15%–17%

Range in value results from variation among studies influenced by climate, soil types, plant growth stage, etc.

Lime prickly-ash is a subdominant species commonly found throughout south Texas and often associated with mesquite, prickly pear and wolfberry. It grows in a variety of soils, from shallow rocky to deeper clays and sand.

Values

White-tailed deer browse on the foliage and young stems. Most birds, including white-winged doves, relish the seeds, and many songbirds nest in the larger shrubs. Additionally, many small animals and reptiles use lime prickly-ash in association with other species for protective cover. It is a food plant for butterfly larvae and a source of nectar for adult butterflies.

Goats utilize the foliage throughout the year and frequently strip the bark in winter.

Historically, lime prickly-ash has been used medicinally in Latin American countries. The bark and leaves are powdered and used as a condiment. The wood produces a yellow dye.

Sapodilla Family   SAPOTACEAE

Coma
Sideroxylon celastrinum

(saffron-plum bumelia, antwood, caimito, coma resimera)

Description

A medium-sized, semi-evergreen spiny shrub or small tree (15'–20') with clustered, dark green, glossy, teardrop-shaped alternate leaves ($^1/_4$"–$^1/_2$" W X <1" L; maybe growing longer on new growing shoots). Between 2 and 10 greenish white flowers are produced from August to November, with the narrowly oblong, black berries produced the following April through June. The bark is mottled-gray-to-brown with long, sharp spines at the ends of the twigs. In late winter it may be the only green tree visible and appears similar to a *live oak tree with thorns*.

Coma grows in various soil types, including in sandy loam, gravelly hills and salt marshes. It commonly grows in mottes within mixed-brush chaparral habitats.

*Crude Protein Value				
	spring	summer	fall	winter
leaves:	14%–20%	13%–16%	13%–15%	12%–16%
fruit:	13%	N/A	N/A	N/A

*Range in value results from variation among studies influenced by climate, soil types, plant growth stage, etc.

Values

The fruit and seeds are eaten by various species of birds, including white-winged doves and chachalacas, and mammals, including raccoons and coyotes, and the leaves are browsed by white-tailed deer. The tree is frequently used for nesting, roosting and loafing by most birds and provides cover for many species of wildlife.

Livestock browse on the leaves and the mottes provide them with shade and protection.

The fruit is eaten in Mexico and is used as an aphrodisiac. The heartwood is occasionally used in cabinet work. Coma may also be used as an ornamental tree.

Amargosa
Castela erecta

(goatbush, allthorn goatbush, bitterbush)

Description

A spiny, multiple-branched, small-to-medium-sized shrub (3'–10') with grayish-white, thorn-tipped branches, and tiny, four-petaled, red-to-pink flowers. The small, simple, alternate leaves (1/4"–1") are linear or oblong, and *very bitter to taste*. The fruit is a tiny, round (1/4"–1/3"), flattened, red drupe with a solitary seed. Amargosa is frequently confused with *lotebush*, but is easily distinguished from the latter by the silvery underside of amargosa leaves.

Amargosa is common throughout south Texas, and is frequently found on gravelly hills and bluffs in thickets and mesquite prairies.

Crude Protein Value				
	spring	summer	fall	winter
leaves:	11%	10%	12%	12%
fruit:	8%	N/A	N/A	N/A

Values

Due to its bitter taste, amargosa is of limited wildlife food value; however, white-tailed deer browse the leaves and eat the fruit. The thorny plants protect small mammals, and birds occasionally nest in them.

The livestock value is unknown and the spines can cause injury.

Historically, amargosa was a popular medicinal plant, with extracts used as remedies for such ailments as intestinal disturbances, fever, skin disease, yellow jaundice and dysentery.

Nightshade Family   SOLANACEAE

Wolfberry
Lycium berlandieri

(desert thorn, Berlandier wolfberry, tomatillo)

Crude Protein Value

	spring	summer	fall	winter
fruit:	N/A	N/A	17%	N/A

Description

A slender, thorny, small shrub (3'-7') with whitish-gray-to-reddish branches, and linear, alternate, clustered leaves (³/₈"-1"). Except for new growth, *2–5 leaves erupt from the same location or are whorled along the stem where the thorn grows*. Wolfberry is a summer deciduous plant, losing its leaves from April to September. It is one of the few shrubs to have leaves during winter. The blue, lavender or white flowers are solitary or may grow in clusters, and the fruit is a small, round, red berry with numerous seeds.

Wolfberry is a common component in mixed-brush chaparral communities on various soil types, ranging from gravelly, rocky, limestone hills to clay flats. It prefers heavier, sandy loams with good drainage, and is often associated with mesquite and prickly pear.

Values

White-tailed deer and occasionally livestock browse on the leaves, and most birds and small mammals, such as chachalacas and raccoons, eat the fruit.

Elm Family   ULMACEAE

Granjeno
Celtis pallida

. . . .

(spiny hackberry, desert hackberry, capul, palo blanco)

Description

A medium-sized evergreen shrub (4'-15') with *strongly zigzagged, smooth, gray branches* and stout, paired thorns. The simple, alternate, green leaves (¹/₂"-2" L X ¹/₂" W) are oblong with toothed margins. It has inconspicuous, greenish-white flowers and small, round (¹/₂"), orange berries.

Granjeno is an important component of mixed-brush chaparral communities and occurs in a variety of soils and habitat types. It is frequently found in thickets, along fence lines and under trees, where birds deposit the seeds.

	spring	summer	fall	winter
leaves:	19%–28%	21%–24%	20%–25%	15%–21%
fruit:	19%–20%	N/A	N/A	N/A

Range in value results from variation among studies influenced by climate, soil types, plant growth stage, etc.

Values

Granjeno is an excellent wildlife food and cover plant. The leaves and stems are browsed by white-tailed deer. The berries provide food for many birds, including white-winged doves, mourning doves, bobwhite quail, scaled quail, cactus wrens, cardinals, pyrrhuloxias, towhees, mockingbirds, thrashers and green jays, and small mammals, including coyotes, raccoons, cottontail rabbits and jackrabbits. The berries are an important water source for birds and small mammals. Granjeno also provides nesting, loafing and roosting sites for birds. Reportedly, verdin prefer granjeno for nesting. Half-cutting and pruning the limbs may provide denser cover for bobwhite and scaled quail. Butterfly larvae feed on the leaves, and the plant yields good flavored honey. Planting granjeno in the backyard in urban areas may improve the site for, and thus benefit, many birds.

Granjeno is eaten by cattle, sheep and goats, especially when availability of other forage is restricted.

The leaves are edible. American Indians ground the fruit and ate it with fat or parched corn. The wood has been used for fence posts and fuel.

Thornless Plants

Amaryllis Family   AGAVACEAE

Spanish Dagger
Yucca treculeana

(yucca, Spanish bayonet, pita, palma loca, palma pita)

Description

A tree-like evergreen plant (3'-10')
with a simple, leafy trunk, and possibly
a few branches at the top, having large,
symmetrical buds of radiating, long, sharp
(sword-like), silver-green or light green
leaves. The white, fragrant, waxy flowers
grow in large clusters at the end of the
stalk, and are generally produced in the
spring and summer of alternate years.
The fruit is a long, cylindrical capsule
(4") packed with numerous black, flat,
triangular seeds.

Spanish dagger is a minor
component, rather thinly scattered,
but commonly found throughout south
Texas in all soil types, habitats and
plant associations.

Values

Deer browse the blooms and, together
with livestock occasionally, also the
leaves. The leafy trunks are browsed by
javelina. Birds such as Inca doves, ground
doves, Harris' hawks and mockingbirds
nest in Spanish dagger. It is a food plant
for butterfly larvae and moths, which
pollinate it.

The flowers are edible if pickled,
boiled or otherwise cooked. Spanish
dagger is grown as a landscape and
ornamental plant in hot, dry areas, rock
gardens and xeriscapes. Historically,
American Indians used the plant as a
source of food and of fiber for ropes,
mats, baskets and sandals; they also made
soap from the roots. The leaves were also
used for thatching huts and as primitive
tools. The seeds are considered to have
laxative properties.

Crude Protein Value

unknown

Littleleaf Sumac
Rhus microphylla

(desert sumac, winged sumac, small-leaved sumac, correosa, scrub sumac)

Description

A clump-forming, densely branched, deciduous shrub (3'-10') with crooked, stiff, smooth, dark branches that become rough with age. The small, dull-green, hairy, odd-compounded leaves (1/2"-1 1/2") have a *winged rachis*. The small, clustered, whitish *flowers appear before the leaves*. The reddish-orange, slightly hairy, clustered, round fruits mature from May to July.

Littleleaf sumac is a minor component in the south Texas brushlands, occurring in a variety of soils but preferring dry, rocky, gravelly, upland mixed-brush thickets, open alkali flats, and mesas and foothills. It is found in association with

Crude Protein Value				
	spring	summer	fall	winter
leaves:	16%	N/A	N/A	N/A

many shrub species, including blackbrush, guajillo and coma.

Values

The fruit is eaten by rodents and birds, including quail and turkeys. White-tailed deer occasionally browse the leaves. Birds nest in this shrub and the dense canopy provides cover for reptiles and small mammals.

There is little browse value for livestock.

The small fruits make a tart drink when crushed in water. Littleleaf sumac is drought tolerant and provides some ornamental, landscape and hedge value, as well as urban and backyard wildlife habitat.

Barberry Family   BERBERIDACEAE

Agarito
Mahonia trifoliata

(desert holly, agarita, algerita, wild currant,
chaparral berry, palo amarillo, paisano bush)

*Crude Protein Value				
	spring	summer	fall	winter
leaves:	13%–20%	N/A	16%	N/A

*Range in value results from variation among studies influenced by climate, soil types, plant growth stage, etc.

Description

An evergreen shrub (3'–8') with 3 stiff, spiny, *holly-like, blue-green leaflets (2"–4") per leaf.* The clustered, yellow, cup-shaped flowers are produced from February to April before the plant bears a small, bright red berry ($^1/_3$"–$^3/_4$"), from April to July. The bark is gray to reddish-brown with yellow wood.

Agarito is a very hardy plant found on dry, rocky slopes and pastures, usually growing independently from other species or in close proximity to trees such as live oaks and mesquite. It prefers neutral-to-alkaline soils and an arid climate. It is a minor component of mixed-brush communities, and is generally more common in the eastern and northern portions of south Texas and rare in central and western portions.

Values

Agarito is a valuable wildlife plant. Young leaves are eaten by deer and other herbivores. The berries are eaten by birds, including quail, cardinals and mockingbirds, and by mammals, including raccoons and opossums, but the hard, spiny leaves make access to them difficult. Many birds and small mammals, including quail and rabbits, utilize agarito for cover, shade and protection. The plant structure also protects other plants growing beneath it from foraging animals. The flowers are a source of nectar for bees and butterflies, and certain moth larvae eat the young leaves.

Goats, sheep and cattle eat the tender young leaves and berries, but the stiff older leaves may injure the mouth.

The young leaves are edible for humans also, and the fruit is frequently used to make a delicious jelly and wine. A yellow dye is made from the wood and roots. Agarito is an attractive, low-maintenance plant often used as an ornamental, landscape or hedge plant and makes a good backyard wildlife plant. Pioneers reportedly used root potions to treat toothaches and dry syphilis sores.

Anaqua
Ehretia anacua

. . . .

(sandpaper tree, knock-away, manzanita, manzanillo)

Description

A subtropical, semi-evergreen, medium-sized tree or shrub (15'–30'+) usually having multiple trunks and a rounded canopy. The olive-to-dark-green leaves (1"–3" L X ³/₄"–1¹/₂" W) are *rough surfaced*, simple, alternate and oval to oblong, and the bark is thick, grooved and gray to reddish brown, with scales often flaking off. The fragrant white flowers are produced in March and April and grow in clusters at the branch ends; they are followed by small,

yellowish-orange, clustered, round berries ($1/4$"–$1/3$").

Anaqua is found in all types of soil but is most common in the Rio Grande Valley. It grows largest in river bottoms and in alkaline soils with good drainage.

Values

White-tailed deer occasionally browse the leaves. The fruit is sweet, highly palatable and is eaten by birds, including quail, white-winged doves and chachalacas, and by mammals, including coyotes, raccoons and feral hogs. Anaqua provides good cover for wildlife, and birds nest, loaf and roost in it. Bees and butterflies are attracted to the flowers, which provide a good nectar source.

Livestock occasionally browse the leaves, which also provide good shade and cover for them.

Anaqua is drought tolerant and is a desirable, attractive ornamental, landscape, and urban wildlife tree often planted for shade in south Texas. It has been used for erosion control along streambeds and hillsides, and the wood has been used to make wheels, fence posts, tool handles, wheel spokes and axles.

Wild Olive
Cordia boissieri

(Mexican olive, anacahuita, chapote prieto)

Description

A stout, *short-trunked, thick-branched*, subtropical evergreen tree (6'-12') with a rounded crown and thick, gray, ridged bark. The large, thick leaves (3"-5" L X 3"-4" W) are soft and velvety, with a hairy, brown underside. The clustered, showy, *trumpet-shaped, tissue-like flowers (2"-3" across) are white with a yellow throat*, and are produced throughout the summer, especially after rains, primarily from April to June. The sweet, whitish fruit (1") matures in July and September.

Wild olive is a minor component that is locally common in drier soils in the southern areas of south Texas, reaching the northern limits, dictated by the plant's inability to tolerate cold, in Jim Hogg, Brooks and lower Duval counties.

Values

The fruit is eaten by many birds and mammals, including deer and javelina. Hummingbirds and bees are attracted to its flowers, and birds nest, roost and loaf in it.

Livestock eat the fruit and utilize the shade during hot summer days.

The fruit has reportedly been used to cure sore throats and coughs, and the leaves to treat rheumatism and bronchial disturbances. Wild olive is drought tolerant and makes an attractive landscape, ornamental and backyard wildlife plant throughout south Texas, as far north as San Antonio.

Desert Yaupon
Schaefferia cuneifolia

(capul, panalero)

Description

A small, densely branched evergreen shrub (3'–6') with smooth gray bark and multiple basal stems. The *teardrop-shaped*, pale green leaves are alternate and clustered, and stay on the tree a long time. The tiny greenish flowers are very inconspicuous and the small, shiny, orange-to-bright-red, roundish fruit grows close to the stem. Desert yaupon is common throughout the western portion of south Texas in various soil types, but prefers heavier soils and rocky hillsides in mixed-brush associations.

*Crude Protein Value				
	spring	summer	fall	winter
leaves:	14%–18%	12%–14%	13%–14%	10%–11%

Range in value results from variation among studies influenced by climate, soil types, plant growth stage, etc.

Values

This drought-tolerant plant is occasionally browsed by white-tailed deer. Birds, including bobwhite quail, scaled quail and cactus wrens, and small mammals, including coyotes and wood rats, eat the fruit. Birds occasionally nest in this shrub.

It provides limited browse for sheep, goats and cattle.

In Mexico, the roots were reportedly used medicinally to cure venereal disease. Due to its evergreen foliage and red berries, it is sometimes used as a landscape, ornamental or hedge plant.

Four-wing Saltbush
Atriplex canescens

(shad-scale, wingscale, chamiso, costillas de vaca)

Description

An erect evergreen shrub, loosely to densely branched (3'–6') with numerous closely attached, alternate leaves ($1/4$"–2" L X $1/8$"–$3/4$" W) that are thick and scaly gray, above and below. The small spikelets of yellowish flowers are produced during the summer and the fruit, usually found in August through October, is easily recognized by its *four-winged shape*.

Crude Protein Value				
	spring	summer	fall	winter
leaves:	N/A	N/A	N/A	12%

Four-wing saltbush is a drought- and heat-tolerant, deep-rooted shrub found in many soil and range types, including dry mesas, salt flats, prairies and hillsides. It is more common in the western than in the eastern half of south Texas.

Values

Four-wing saltbush is a preferred, palatable and nutritious wildlife plant, especially during winter and times of drought. The foliage is browsed by deer and rabbits and the seeds are eaten by birds, including quail and songbirds, and by small mammals, including squirrels, ground squirrels, mice, rats, rabbits and porcupines.

It is a nutritious feed for cattle, sheep and goats, but concentrated feeding can cause digestive problems in livestock, especially during droughts and cold weather.

Four-wing saltbush is useful in erosion control because it is deep rooted. It has reportedly been used in immunization extracts for hay fever. It has ornamental value, particularly in drier areas.

Persimmon Family   EBENACEAE

Texas Persimmon
Diospyros texana

(Mexican persimmon, chapote)

Description

A semi-evergreen shrub or small tree (6'–15') with a crooked trunk, densely tangled limbs, *smooth gray flaking bark and alternate, leathery, dark green leaves (1"–2") which curl on the margins*. The bell-shaped, sweet, fragrant flowers are small and greenish white and the round fruit (1") is dark purple to black when ripe.

Texas persimmon is common throughout south Texas. The shrub frequently forms mottes and is associated with most mixed-brush communities. It is found in various soil types, including sandy, shallow and rocky sites.

*Crude Protein Value

	spring	summer	fall	winter
leaves:	18%	6%–14%	12%	10%

Range in value results from variation among studies influenced by climate, soil types, plant growth stage, etc.

Values

Texas persimmon is a valuable wildlife plant. Deer browse the leaves and the sweet, juicy fruit is eaten by birds, including quail, turkeys and songbirds, and by mammals, including deer, javelina, feral hogs, coyotes, raccoons, possums and skunks. Birds nest, roost and loaf in the trees, which also provide shade and cover for small mammals. The flowers attract many pollinators, such as bees, and the plant is a food source for butterfly larvae as well as a source of nectar for adult butterflies.

Texas persimmon is occasionally browsed by cattle, but is of low preference for sheep and goats.

The fruit is edible. In Mexico, it is reportedly used to make a black dye, while the wood is used to make tools and engraving blocks. It is an excellent ornamental, landscape and urban wildlife tree, especially when space permits groves.

Vine Ephedra
Ephedra antisyphilitica

(Mormon tea, clapweed, popote, canatilla, popotillo)

Description

A small, erect or low-spreading, *leafless-appearing* shrub (1'–4') with stiffish, green, multi-noded stems, that are opposite or whorled branching at the nodes. The tiny, scale-like leaves are cone-like, with a narrow, tannish band circling the stem at their base. Male and female plants are separate. The tiny cones are green, yellowish, or reddish and the small fruit is smooth, solitary, succulent and reddish.

Vine ephedra is most often found in gravelly, rocky limestone hillsides or fields, and in arroyos, canyons and ravines in association with most mixed-brush species.

Values

Vine ephedra is a preferred browse plant of white-tailed deer and cattle. It provides cover and/or nesting for several species of small mammals and reptiles.

Anglo settlers reportedly boiled this plant to make a tea used as a preventative against syphilitic infection. American Indians still regularly use ephedra as a tea for any kind of urinary disturbance. The shrub is sometimes used as an ornamental plant, and also has some urban wildlife value.

Spurge Family   EUPHORBIACEAE

Southwest Bernardia
Bernardia myricifolia

(myrtle-croton, oreja de raton)

*Crude Protein Value

	spring	summer	fall	winter
leaves:	15%-20%	N/A	N/A	N/A

*Range in value results from variation among studies influenced by climate, soil types, plant growth stage, etc.

Description

A small, densely branched shrub (3'-8') having simple, alternate, dark green leaves ($1/2$"-2") with a wavy margin and lighter and densely hairy undersides. The flowers are small and inconspicuous, and the rounded, 3-seeded fruit ($1/3$"-$1/2$") is grayish brown.

Southwest bernardia is a minor component of mixed-brush communities, but is fairly common throughout the Rio Grande Plains, and is found on dry caliche and rocky slopes, hills and canyons. It is commonly associated with cenizo, Texas kidneywood, guajillo and blackbrush.

Values

Southwest bernardia is drought resistant and palatable, and is browsed by deer and livestock. Seed-eating birds such as quail, doves, cardinals and sparrows readily eat the seeds. It is a food plant for butterfly larvae and a source of nectar for adult butterflies.

Mountain Laurel
Sophora secundiflora

(mescalbean, coral bean, big-drunk bean, frijolito)

Description

A multiple-based evergreen shrub or small tree (3'-12') with dark-gray-to-black, rigid, upright branches and velvety young stems. The dark, glossy, *shiny green leaves* are oddly compound, oblong and somewhat leathery, with 5-11 leaflets per leaf. The large, fragrant, showy, bluish-purple flowers are grouped on elongated clusters and are produced in March and April. The *dark, woody, thick, fairly large legumes* (1"-5") are brown, aromatic, and semi-segmented.

Mountain laurel is found throughout south Texas, most often on shallow, gravelly or limestone soils in various plant associations.

Values

The seeds and leaves are toxic to humans, livestock and wildlife. There is very little known wildlife value. Bees and butterflies are attracted to the flowers, which provide a good nectar source.

The seeds and leaves are toxic to cattle and, to a lesser extent, to sheep and goats.

A yellow dye can be made from the wood and the red seeds have been used as beads for necklaces and other jewelry. Mountain laurel is drought tolerant and is frequently used in landscaping as an ornamental shrub, but it is slow growing.

Crude Protein Value

unknown

Legume Family   FABACEAE

Texas Kidneywood
Eysenhardtia texana

(vara dulce, rockbush)

Description

An irregularly shaped, multiple-based evergreen shrub (3'-8') with slender gray stems. The leaflets (15–47) are tiny, and form an alternately, odd-compounded red leaf. The tiny flowers are white and hairy. They are borne on elongated clusters (1"–4") blooming between April and November, especially after rains. The fruit is a small, dotted, thickened, green-to-brown legume, with 2–4 seeds. *Crushing the leaves between your fingers emits a strong pungent odor.*

Texas kidneywood is found throughout south Texas, but more commonly in the northern areas, frequently on calcareous soils and dry limestone rocky hills and canyons, in mixed-brush associations.

	spring	summer	fall	winter
leaves:	24%–26%	20%–22%	17%–23%	17%–20%

Range in value results from variation among studies influenced by climate, soil types, plant growth stage, etc.

Values

Texas kidneywood is highly preferred for browsing by white-tailed deer; livestock also browse on it. Birds eat the seeds, and the flowers provide a source of honey and butterfly nectar.

It has reportedly been used to make orange dyes, and also medicinally as a diuretic. Texas kidneywood can be used as an ornamental and landscape shrub.

Live Oak
Quercus virginiana

(encino)

Description

A small shrub, becoming a large evergreen tree (10'-50') capable of forming a large, spreading canopy, with heavy limbs which occasionally touch the ground. The simple, alternate leaves (1"-4" L X 1/2"-1" W) are thick, shiny and dark green on the top and lighter underneath, with toothed margins curling downward. The male flowers are produced from March to May on catkins (2"-3") that appear yellowish and hairy.

The bark is gray or black and somewhat grooved, and the fruit is an acorn (1").

Live oak is commonly found throughout the northern and eastern portions of south Texas, usually in sandy or gravelly clay soils, often growing in mottes. It is also found along major rivers and drainages, but seldom grows in upland sites dominated by mixed-brush chaparral.

	spring	summer	fall	winter
leaves:	13%	10%	11%	9%
acorns:	N/A	N/A	N/A	6%

*Range in value results from variation among studies influenced by climate, soil types, plant growth stage, etc.

Values

Live oak is a valuable plant for wildlife, for food and cover. New foliage is browsed by white-tailed deer; many birds, including turkeys, as well as mammals, including deer, feral hogs, javelina and squirrels, eat the acorns. Birds, including songbirds, turkeys and raptors, and some mammals, including squirrels, use the tree for nesting and/or roosting.

Cattle, sheep and goats browse the new leaves and eat the acorns. The tree provides excellent shade for livestock.

Live oak is frequently used as an ornamental or landscape tree on account of its aesthetic value and shade. The hard, strong wood was used historically for ship building and furniture. It also makes excellent firewood and the bark is used for tanning hides into leather. American Indians reportedly used acorn oil for cooking, and acorn meal to make bread-like patties.

Mint Family   LAMIACEAE

Shrubby Blue Sage
Salvia ballotiflora

(blue sage, shrubby blue salvia, mejorana)

Description

A much-branched, aromatic shrub (2'–6') with pale, dark gray, *square* stems. The hairy, grayish-colored leaves ($1/2$"–$1 1/2$") are simple, opposite and have serrated margins. The bluish-purple, triangular, 3-lobed flowers are found in elongated clusters blooming intermittently all summer. The fruit is very tiny and inconspicuous.

Shrubby blue salvia is a minor component of mixed-brush chaparral thickets and is fairly common throughout south Texas. It prefers shallow, rocky, sandy, gravelly or limestone hillsides or brushy slopes.

Values

Deer browse the leaves, and small mammals and reptiles utilize it for cover.

The dried, aromatic leaves are reportedly used for flavoring meats and other foods.

Olive Family   OLEACEAE

Narrowleaf Forestiera
Forestiera angustifolia

(elbowbush, tanglewood, desert olive,
panalero, chaparral blanco)

Description

A stiff, intricately branched, rounded evergreen shrub (3'–8') with smooth, gray branches coming off the main stem *opposite, at 90-degree angles, to form elbows*. The simple, smooth, light green leaves are linear with entire margins and are often clustered on short, knotty spurs of older twigs. The bark and twigs are gray and smooth. The inconspicuous, greenish-yellow flowers are produced in clusters in the spring or before the new leaves erupt, and the fruit is a small one-seeded drupe, purple-black to black in color.

	spring	summer	fall	winter
leaves:	13%–21%	8%–11%	6%–8%	N/A

Range in value results from variation among studies influenced by climate, soil types, plant growth stage, etc.

Narrowleaf forestiera is a minor component that is fairly common in mixed-brush or open woodlands, and is usually found on dry, well-drained hillsides and arroyos in full sun. It is also found along the coast in shelly areas.

Values

White-tailed deer and livestock browse the foliage and the fruit is eaten by many mammals, such as raccoons, foxes, ringtails, rabbits, ground squirrels, rats and mice, and by many birds, such as bobwhite and scaled quail, white-winged doves and numerous songbirds. The plant offers some canopy cover from predators and is an important source of food for bees.

It can be used as a ornamental shrub, especially in saline soils and in windy areas.

Coyotillo
Karwinskia humboldtiana

(Humboldt coyotillo)

Description

A low, conspicuously thornless evergreen shrub (2'-6') with *dark green, strongly veined oblong leaves* ($3/4$"-$1^3/4$" L X $1/2$"-$3/4$" W) and smooth, gray bark. The small flowers are greenish and the small brown or black fruit ripens in late summer and fall.

Coyotillo is a common component found throughout south Texas in all types of soils and habitats, but prefers drier

areas and shallow soils. It frequently grows in association with guajillo and blackbrush communities.

Values

Coyotillo is a toxic plant with little value for wildlife. The seeds are extremely toxic and reportedly affect the nervous system, causing paralysis in the limbs of humans and domestic animals. Some indigenous mammals and birds, such as coyotes and chachalacas, eat the fruit.

The seeds and leaves are poisonous to cattle, sheep, goats, horses and swine, but the plant is usually not grazed unless extreme drought conditions exist.

In Mexico, historically, a medicine was reportedly made from the plant to treat fever. Coyotillo can be used as an ornamental and landscape shrub, especially in dry areas, rock gardens and xeriscapes.

Hogplum
Colubrina texensis

(Texas colubrina)

Description

A low, rounded deciduous shrub (3'-6') with *light gray, short, stiff, zigzag branches and spreading twigs*, and frequently forming dense thickets. The simple, grayish-green leaves (¹/₂"-1") have 3 prominent veins. The inconspicuous, *greenish flowers are star shaped*, and are produced in April and May. The fruit, dark brown to black in color, remains on the plant a long time.

Hogplum is a common component of south Texas brushlands that is found in a variety of mixed-brush habitats and soil types, though it prefers deep, sandy, drier sites. It is an aggressive invader on sites that have been mechanically manipulated.

*Crude Protein Value

	spring	summer	fall	winter
leaves:	18%–24%	15%–19%	15%–22%	17%

*Range in value results from variation among studies influenced by climate, soil types, plant growth stage, etc.

Values

Hogplum is browsed by white-tailed deer, and the fruit and seeds are eaten by birds and small mammals, including javelina. Birds occasionally nest in it. The thicket provides cover for birds, small mammals and reptiles.

Hogplum offers limited browse for livestock, and the seeds are reportedly toxic to sheep.

Cenizo
Leucophyllum frutescens

(purple sage, Texas sage, Texas silverleaf, Texas ranger, barometer bush, ash bush, wild lilac)

Description

A low-growing, rounded evergreen (ever-gray) shrub (3'–6'), with stout stems and *conspicuously light, silvery gray, fuzzy leaves*. The simple leaves are alternate and opposite or whorled, and the flowers, colored pale violet, purple, pink, or sometimes white, are produced shortly after rains. The bright-colored flowers contrast noticeably with the silvery gray leaves. The small fruit has numerous seeds in two sections.

Cenizo is a common component throughout south Texas, but prefers upland sites with shallow soils, such as gravelly or limestone hills, bluffs and ravines. It is frequently found in mixed-brush chaparral, often in association with guajillo, blackbrush and twisted acacia.

*Crude Protein Value				
	spring	summer	fall	winter
leaves:	15%-16%	12%-13%	13%-15%	12%-13%

*Range in value results from variation among studies influenced by climate, soil types, plant growth stage, etc.

Values

White-tailed deer browse cenizo, especially in winter and in drought conditions, and birds occasionally nest in it. Cenizo is a food plant for butterfly larvae and a source of nectar for adult butterflies.

Cenizo is also browsed by cattle, especially when grass is dormant.

This beautiful shrub is often utilized as an ornamental or landscape plant in south Texas. It has reportedly been used by American Indians for the treatment of chills and fever.

Elm Family   ULMACEAE

Cedar Elm
Ulmus crassifolia

(scrub elm, lime elm, Texas elm, basket elm,
southern rock elm, olmo)

Description

A deciduous, slender-crowned tree (30'–60') having rough bark and branches, and *opposite, corky wings on the twigs*. The small, rough, dark green leaves are simple and alternate with *serrated edges* ($1/2$"–2" L) that turn yellow to gold in the fall. The small, red-to-green petal-less flowers are in small

Crude Protein Value

	spring	summer	fall	winter
fruit:	N/A	N/A	6%	N/A

clusters produced in July and August. The small, greenish fruit ($^1/_4$"–$^1/_2$") is flattish round, hairy and winged.

Cedar elm is commonly found along rivers, creeks and drainages and is frequently associated with hackberries, oaks and granjeno in various soil types.

Values

White-tailed deer browse the foliage and the fruits are eaten by many mammals, including squirrels and mice, and by birds, including turkeys. Many birds and small mammals nest, loaf and roost in it. It is a food plant for butterfly larvae.

Livestock browse the foliage, frequently forming a browse line; the tree also provides livestock with shade.

Cedar elm can be used as an ornamental shade or landscape tree. It also provides habitat for some kinds of urban wildlife such as squirrels and birds.

Elm Family   ULMACEAE

Sugar Hackberry
Celtis laevigata

(Texas sugarberry, palo blanco)

Description

A deciduous tree (15'-50'+) with *thin, gray, noticeably warty bark* and a broad crown. The simple, alternate, oblong lanceolate, light green leaves (2$^{1}/_{2}$"-4" L X 1"-2$^{1}/_{2}$" W) have 3 prominent basal veins. The small, inconspicuous, greenish flowers are produced in the spring and the fruit is a small ($^{1}/_{4}$"-$^{1}/_{2}$") brown or orange drupe on a long pedicel, ripening in the late summer.

Crude Protein Value				
	spring	summer	fall	winter
leaves:	28%	24%	25%	19%
fruit:	16%	11%	12%	N/A

Sugar hackberry is widespread throughout south Texas along rivers, creeks, drainages and lowlands, and is frequently associated with other large trees, such as live oaks, mature mesquites and cedar elm. Sugar hackberry is uncommon in upland sites dominated by mixed-brush chaparral. A similar plant is netleaf hackberry, which can be distinguished by its rougher leaves.

Values

Sugar hackberry is a very valuable and desirable wildlife plant. White-tailed deer browse on the foliage, and the fruits and seeds are eaten by many birds, including cedar waxwings, cardinals, robins, mockingbirds, yellow-bellied sapsuckers, turkeys and chachalacas, and also by many mammals, including raccoons. The fruit stays on long after the leaves fall off, thus increasing food availability. It is also a food plant for butterfly larvae. Additionally, the trees provide nesting, roosting and loafing areas for birds.

Sugar hackberry is browsed by livestock and provides valuable shade on hot, dry days.

The wood is occasionally used for flooring, furniture, sporting goods and fuel. It is a relatively fast-growing tree and can be used in shelterbelts and fence rows and for landscaping.

Lantana
*Lantana urticoides**

(calico bush, bunchberry, mejorana,
hierba de cristo, monte cristo)

Description

A deciduous, wide-spreading, aromatic perennial shrub (1'-6') with multiple upright, green-to-reddish, rough and somewhat prickly stems. The younger stems are somewhat squarish. The simple, opposite, *aromatic*, hairy leaves have dark green upper sides and light green undersides, and serrated edges. The showy flowers are *clumped into rounded heads and vary in color* between red, orange and yellow. The flowers are produced throughout the summer from the upper leaf axil. The fruit is a small, dark blue or black drupe, ripening on long peduncles (1"-4") from August to September.

Lantana is a common plant usually found in all types of habitats, but it prefers sandy and gravelly soils and hot, dry areas. It is frequently found in mixed-brush communities, and in fallow fields, along roadsides and in fence rows.

*Referred to as *Lantana horrida* in some sources.

Crude Protein Value

	spring	summer	fall	winter
leaves:	19%	N/A	19%	19%

Values

Lantana is reportedly toxic, with little browse value for wildlife, although some birds, including quail, eat the fruit. Small mammals and reptiles occasionally use the plant for cover. It is a food plant for butterfly larvae and a source of nectar for adult butterflies.

It is also toxic to livestock and humans.

Lantana is an attractive, low-spreading ornamental and landscaping plant. In Mexico, crushed leaves were reportedly used medicinally to treat stomach ailments and snake bites.

Whitebrush
Aloysia gratissima

(bee brush, vara dulce, palo amarillo, bee blossom)

Description

A deciduous, very slender, densely branched, thicket-forming, upright shrub (4'-8') with stiff, squarish, brittle, grayish branches that reveal a yellow wood. The narrowly oblong and lanceolate, dark green, hairy leaves vary in length ($^1/8$"-$1^1/4$"). The small, white, vanilla-scented flowers are produced intermittently from March to November and grow in densely clustered spikelets (1"-3").

Whitebrush is an aggressive invader commonly found in drainages and fertile lowlands, where it forms pure thickets. In upland areas, it associates with most other south Texas brush species, and is found in various soil types and on gravelly hillsides and limestone bluffs.

	spring	summer	fall	winter
leaves:	23%	N/A	19%	22%

Values

Although palatable to deer, whitebrush has limited browse value, and the seeds are not readily eaten by birds. Its value to wildlife is in providing a protective overstory and escape cover for birds, mammals and reptiles, including quail, javelina, feral hogs and bobcats.

Whitebrush, especially the regrowth, is occasionally grazed by cattle and goats, particularly during times of stress. It is toxic to horses, mules and burros.

Bees make a delicious light-colored honey from the flowers. In Mexico, the leaves and flowers have reportedly been used medicinally to treat diseases of the urinary tract. The shrub can be used as a ornamental or landscape plant.

Caltrop Family   ZYGOPHYLLACEAE

Creosotebush
Larrea tridentata

. . . .

(greasewood, gobernadora)

Description

A *creosote smelling*, evergreen shrub (2'-6') with dark nodes on slender, dark gray, multiple-based stems. The leaves have two small, oblong leaflets (<$^{1}/_{2}$"), which are *resinously sticky and strong scented*, and dark green to yellowish green in color. The small, silky, yellow flowers are produced from April to August, after which a small, round, whitish, hairy fruit appears.

Creosotebush is a common desert shrub that is indicative of shallow, poor soil and abused rangeland in hot, dry areas. It is frequently found in the western portion of south Texas, and is usually not closely associated with other plant species.

Values

Creosotebush has no browse value; however, many small mammals, including ground squirrels, eat the seeds and fruit and utilize the plant for shade and cover. Hummingbirds obtain the nectar from the flowers.

88

Crude Protein Value				
	spring	summer	fall	winter
tips:	13%	N/A	N/A	N/A
leaves:	12%	N/A	N/A	N/A

Creosotebush is toxic to sheep, and cattle will not graze on it. An edible livestock feed has been developed from it, however, and a valuable antioxidant has been commercially extracted from it.

American Indians reportedly pickled the buds in vinegar and ate them. Historically, leaf extractions were used medicinally as an antiseptic and to treat rheumatism, venereal disease, tuberculosis and intestinal disorders. When boiled with lard, the leaves make a liniment for cuts and bruises.

This drought-tolerant plant, with its fragrant flowers, can be used as an ornamental or landscape plant, especially when mixed with other southwest desert plants in xeriscapes and rock gardens.

Guayacan
Guajacum angustifolium

(Texas porliera, soapbush, ironwood)

Description

A stout evergreen shrub or small tree (2'-8') with short, irregular, knotty branches that *appear as if the leaves are growing directly from the stems*. The thick, leathery, dark green, opposite leaves are crowded at the nodes, compounded with 4-8 pairs of leaflets (<1") and fold inward in the heat of the day. The fragrant flowers (1" W) are violet or purple, with noticeable yellow anthers, and are frequently found in small clusters produced in March

*Crude Protein Value				
	spring	summer	fall	winter
leaves:	17%–26%	15%–23%	16%–19%	14%–17%

*Range in value results from variation among studies influenced by climate, soil types, plant growth stage, etc.

and April. The fruit is somewhat heart shaped with winged margins, and has large, bean-like, reddish seeds.

Guayacan is a minor component commonly found in mixed-brush chaparral throughout south Texas in most habitats and soil types.

Values

Guayacan is browsed by white-tailed deer and is a limited food source for birds. Small mammals such as rats and mice occasionally use the shrub for cover, as do birds, which additionally use it for nesting and roosting. The flowers are a good source of honey.

Guayacan is grazed by sheep and goats.

Its wood is the hardest in Texas and, indeed, in the entire country. It is often used for fence posts and tool handles, as well as for barbecue wood. Soap can be made from the bark of the root, and root extracts are reportedly used to treat rheumatism and venereal disease. The plant is drought tolerant and can be used as a hedge, landscape or ornamental shrub, especially in rock gardens and xeriscapes.

Nutritional Values of Plants*

PLANT	CRUDE PROTEIN Spring	Summer	Fall	Winter	season not indicated	DIGESTIBLE PROTEIN Spring	Summer	Fall	Winter	season not indicated	DIGESTIBLE DRY MATTER Spring	Summer	Fall	Winter	SOURCE
Agarito	13–16	14.0	16.0 10.1												Huston, et al., 1981 NRCS, 1990 White, 1979
Amargosa mast+	10.9	10.4	11.6	11.7						7.5	53.0	59.0	59.0	60.0	Everitt & Gonzalez, 1981 Everitt & Alaniz, 1981
Anaqua										8.9					Everitt & Alaniz, 1981
Blackbrush	18.2 15.0 20.0	17.5 15.0	19.8 15.0 16.0 12.1	16.5 14.0							22.4 34.0	20.5 29.0	26.2 37.0	27.8 26.0	Lynch, 1977 Meyer et al., 1984 Richardson, C.L., 1990 White, 1979
Brasil	23.8 13.0	14.3	17.1	17.5							51.7	38.7	34.7	41.6	Lynch, 1977 Meyer et al., 1984
mast+	21.0	17.0	18.0	16.0						7.9	60.0	48.0	55.0	50.0	Richardson, C.L., 1990 Everitt & Alaniz, 1981
Catclaw	21.1 21–30	16.1 17–19 23.3 18.0	16.0 18.5 19.0 13.0	17.0 17.0 13.0							53.0 46.6 61.0	45.0 40.6 50.0	37.0 33.9 53.0	34.7 47.0	Everitt & Gonzalez, 1981 Huston, et al., 1981 Lynch, 1977 Richardson, C.L., 1990 NRCS, 1990 White, 1979
	24.0														
Cedar elm	No values reported														
Cenizo	14.7 16.0	11.6 13.0	12.5 15.0 10.7	12.5 12.0							63.0 57.0	55.0 50.0	49.0 55.0	51.0 50.0	Everitt & Gonzalez, 1981 Richardson, C.L., 1990 White, 1979

*All values are expressed as percentages of the total plant matter.
+Mast includes fruit, beans and nuts.

Coma	14.1 17.7 20.0	13.4 15.9 16.0	13.3 15.1 15.0	11.9 15.7 15.0						49.0 49.3 51.0	50.0 35.7 47.0	44.0 32.4 48.0	48.0 38.2 40.0	Everitt & Gonzalez, 1981 Lynch, 1977 Richardson, C.L., 1990
Coyotillo	No values reported													
Creosotebush				12.8	12.8									Meyer & Karasov, 1989
Ebony mast+	23.1	20.1	22.6	20.7		22.18 (seed only)				57.0	48.0	45.0	46.0	Everitt & Gonzalez, 1981 Everitt & Alaniz, 1981
Four-wing saltbush leaves stems	20.3 23.7 14.7	15.1 7.9	18.9 8.4	12.0	16.6									Gabel, 1990 NRCS, 1981 Garza & Fulbright, 1988 Garza & Fulbright, 1988
Granjeno mast+	22.4 28.3 19.0 28.0	20.8 23.5 22.0	19.8 24.5 22.0 23.0	15.2 19.0 19.0		19.9				67.0 64.3 72.0	67.0 67.0 67.0	56.0 65.0 69.0	63.0 66.3 67.0	Everitt & Gonzalez, 1981 Lynch, 1977 Meyer et al., 1984 Richardson, C.L., 1990 Everitt & Alaniz, 1981
Guajillo mast+	27.7 27.0 20.0	21.4 20.0 15.6–20.2	22.2 21.0 16.8	21.4 17.0 17.6	9.1	2.14–7.31	3.5	4.0	17.07 (seed only)	38.3 48.0	26.8 40.0	30.9 47.0	29.3 43.0	Lynch, 1977 Richardson, C.L., 1990 Barnes et al., 1991 Everitt & Alaniz, 1981
Guayacan	17.9 26.1 21.0	16.6 22.6 17.0	17.4 18.8 18.0	15.0 17.4 16.0						47.0 45.3 58.0	51.0 41.2 57.0	51.0 43.4 58.0	50.0 46.2 55.0	Everitt & Gonzalez, 1981 Lynch, 1977 Richardson, C.L., 1990
Hogplum	17.8 24.0	15.0 19.0	15.4 22.0 12.8	17.0						56.0 59.0	49.0 50.0	49.0 54.0	50.0	Everitt & Gonzalez, 1981 Richardson, C.L., 1990 White, 1979
Huisache mast+	23.0					17.6								Meyer et al., 1984 Everitt & Alaniz, 1981

+Mast includes fruit, beans and nuts.

continued

PLANT	CRUDE PROTEIN					DIGESTIBLE PROTEIN					DIGESTIBLE DRY MATTER					SOURCE
	Spring	Summer	Fall	Winter	season not indicated	Spring	Summer	Fall	Winter	season not indicated	Spring	Summer	Fall	Winter	season not indicated	
Kidneywood	24.2	20.4	17.1	17.0							53.1	45.9	53.0	45.2		Lynch, 1977
	26.0	22.0	23.0	20.0							62.0	57.0	50.0	54.0		Richardson, C.L., 1990
			10.9													White, 1979
Lantana	No values reported															
Lime prickly-ash	17.1	15.6	16.6	15.8							63.0	75.0	71.0	70.0		Everitt & Gonzalez, 1981
	21.1	15.9	18.5	16.9							51.5	60.6	48.3	64.5		Lynch, 1977
			18.0													Meyer et al., 1984
	21.0	16.0	17.0	15.0							67.0	58.0	65.0	62.0		Richardson, C.L., 1990
Littleleaf sumac	16.0	10.0														NRCS, 1990
		11.6														NRCS, 1990
																NRCS, 1989
	16.1															TPWD, 1994
Live oak	10-20	9-10	10-12	9-10							57.0	49.0	51.0	48.0		Huston et al., 1981
	13	10.0	11.0	9.0												Richardson, C.L., 1990
			7.5													White, 1979
																Everitt & Alaniz, 1981
																Richardson, C.L., 1990
mast+										5.7						
mast+										6.0						
Lotebush	18.5	14.9	16.3	11.7							59.0	52.0	32.4	30.0		Everitt & Gonzalez, 1981
	18.0	16.7	20.0	15.0							38.1	33.8	44.0	39.0		Lynch, 1977
	24.0	19.0									51.0	48.0				Richardson, C.L., 1990
Mesquite (dry dormant leaf only)	26-32			16.0												NRCS, 1990
mast+		16.0										62.0				Huston et al., 1981
mast+		13.0														Richardson, C.L., 1990
mast+		13.0														NRCS, 1990
mast+			8.8													White, 1979
mast+		11.2	12.1									59.0				Everitt & Gonzalez, 1981
mast+										12.5						Everitt & Alaniz, 1981
Mountain laurel		17-18								12.0						Huston et al., 1981
Narrowleaf forestiera																
mast+										6.7						Everitt & Alaniz, 1981

+Mast includes fruit, beans and nuts.

Table continued.

Plant										Reference
Palo verde	24.3									TPWD, 1995
Prickly pear	8.5	6.0	6.6	6.2			76.0	80.0	78.0	Everitt & Gonzalez, 1981
	2–7	7.0								Huston et al., 1981
	13.3	5.6	10.3	2–3						Lynch, 1977
	12	7.0	9.0	5.5		68.4	66.6	66.7	62.6	Richardson, C.L., 1990
mast+		6.0	8.0	5.0		75.0	69.0	71.0	68.0	Richardson, C.L., 1990
mast+		6.2	8.3					63.0		Everitt & Gonzalez, 1981
mast+		2.0			7.1		73.0	58.0		Everitt & Alaniz, 1981
mast+ *(pulp only)*										NRCS, 1990
Retama	20.1									TPWD, 1995
Shrubby blue sage	12.5	14.3	14.4	10.7	9.9	52.1	47.7	47.8	48.4	TPWD, 1994
	18.2									Lynch, 1977
mast+										Everitt & Alaniz, 1981
Southwest bernardia	14.9									NRCS, 1989
	20.1									TPWD, 1994
Spanish dagger *(flowers only)*	14–22		7.0							Huston et al., 1981
Sugar hackberry					10.7					Everitt & Alaniz, 1981
mast+										NRCS, 1990
new leaf/twig	20.0									NRCS, 1990
mature leaf/twig	8.0									
Tasajillo	8.3	8.0			7.6	63.0	62.0			Everitt & Gonzalez, 1981
mast+										Everitt & Alaniz, 1981
Texas persimmon	14–25	10–12	9–11	10.0		58.0	51.0	58.0	41.0	Huston et al, 1981
	18.0	14.0	12.0							Richardson, C.L., 1990
			10.7							White, 1979
mast+	6.0									Richardson, C.L., 1990
mast+	9.85 *(pulp only)*									Everitt & Alaniz, 1981
Twisted acacia	16.9	19.6	21.6	16.7		28.4	27.0	28.6	28.9	Lynch, 1977
	22.0	18.0	20.0	16.0		39.0	37.0	33.0	28.0	Richardson, C.L., 1990
mast+		10.0								Richardson, C.L., 1990

+Mast includes fruit, beans and nuts.

continued

PLANT	CRUDE PROTEIN					DIGESTIBLE PROTEIN					DIGESTIBLE DRY MATTER				SOURCE
	Spring	Summer	Fall	Winter	season not indicated	Spring	Summer	Fall	Winter	season not indicated	Spring	Summer	Fall	Winter	
Vine ephedra	12.3 16.4 11.6	11.9 14.5	12.9 17.8	11.8 14.6							59.0 56.6	55.0 46.8	51.0 53.2	48.0 47.1	Everitt & Gonzalez, 1981 Lynch, 1977 NRCS, 1989
Whitebrush	23.0	19.0	22.0 13.6								58.0	51.0	55.0		Richardson, C.L., 1990 White, 1979
Wild olive	No values reported														
Wolfberry mast+										17.3					Everitt & Alaniz, 1981

+Mast includes fruit, beans and nuts.

Bibliography

Ajilvsgi, Geyata. 1979. *Wildflowers of the Big Thicket, East Texas and Western Louisiana.* Texas A&M Univ. Press: College Station. 360 pp.

Arnold, L.A., Jr., and D.L. Drawe. 1979. "Seasonal food habits of white-tailed deer in the south Texas plains." *J. Range Manage.* 32(3):175–178.

Barnes, T.G., L.H. Blankenship, L.W. Varner, and J.F. Gallagher. 1991. "Digestibility of guajillo for white-tailed deer." *J. Range Manage.* 44:606–610.

Beasom, S.L., J.M. Inglis, and C.J. Scrifes. 1982. "Vegetation and white-tailed deer responses to herbicide treatment of a mesquite drainage habitat type." *J. Range Manage.* 35(6):790–794.

Bryant, F.C., F.S. Guthery, and W. Webb. 1981. "Grazing management in Texas and its impact on selected wildlife." *In:* L. Nelson, Jr. and J.M. Peek, eds. *The Proceedings of Wildlife–Livestock Relationships*, pp. 94–112. Symposium publ. Forest, Wildlife, and Range Experiment Station: Moscow, ID.

Bryant, F.C., C.A. Taylor, and L.B. Merrill. 1981. "White-tailed deer diets from pastures in excellent and poor range condition." *J. Range Manage.* 34(3):193–199.

Correll, D.S., and M.C. Johnston. 1970. *Manual of the Vascular Plants of Texas.* Univ. of Texas: Dallas. 188 pp.

Cox, P.W., and P. Leslie. 1988. *Texas Trees: A Friendly Guide.* Corona Publishing Company: San Antonio. 374 pp.

Crosswhite, F.S. 1980. "Dry country plants of the south Texas plains." *Desert Plants.* 2(3):141–179.

Davis, C.E. 1990. *Deer management in the south Texas plains.* Fed. Aid Project W125R. Texas Parks and Wildlife Department: Austin. 32 pp.

Davis, C.E., and L.L. Weishuhn. 1982. *South Texas deer-livestock relationships and management.* Fed. Aid Project W109R. Booklet 7000-60. Texas Parks and Wildlife Department: Austin. 10 pp.

Davis, R.B., and C.K. Winkler. 1968. "Brush vs. cleared range as deer habitat in southern Texas." *J. Wildl. Mgt.* 32(2):321–329.

Davis, W.B., and D.J. Schmidly. 1994. *The Mammals of Texas.* Texas Parks and Wildlife Press: Austin. 338 pp.

Dodd, J.D. 1968. "Mechanical control of prickly pear and other woody species in the Rio Grande Plains." *J. Range Manage.* 21:366–370.

Drawe, D.L. 1968. "Mid-summer diet of deer on the Welder Wildlife refuge." *J. Range Manage.* 21(3):164–166.

Drawe, D.L., and I. Higginbotham, Jr. 1980. "Plant communities of the Zachary ranch in the south Texas plains." *Texas J. of Science.* 32(4):319–332.

Everitt, J.H., and M.A. Alaniz. 1981. "The nutrient content of cactus and woody plant fruits eaten by birds and mammals in south Texas." *Southwestern Nat.* 26(3):301-305.

Everitt, J.H., and D.L. Drawe. 1993. *Trees, Shrubs, and Cacti of South Texas.* Texas Tech Univ. Press: Lubbock. 214 pp.

Everitt, J.H., and C.L. Gonzalez. 1981. "Seasonal nutrient content of food plants of white-tailed deer in the south Texas plains." *J. Range Manage.* 34(6):506-510.

Forbes, T.D.A., I.J. Pemberton, G.R. Smith, and C.M. Hensarling. 1995. "Seasonal variation of two phenolic amines in *Acacia berlandieri. J. Arid Env.* 30:403-415.

Fulbright, T.E. 1987. "Effect of repeated shredding on a guajillo (*Acacia berlandieri*) community." *Texas J. of Agriculture & Nat. Resources.* 1:32-33.

Fulbright, T.E., and S.L. Beasom. 1987. "Long term effects of mechanical treatments on white-tailed deer browse." *Wildl. Soc. Bull.* 15:560-564.

Fulbright T.E., J.P. Reynolds, S.L. Beasom, and S. Demaris. 1991. "Mineral content of guajillo regrowth following roller chopping." *J. Range Manage.* 44:520-522.

Gabel, R. 1990. "Four winged saltbush merits to cattle production." *The Cattleman.* April 1990:76-80.

Garza, A., Jr., and T.E. Fulbright. 1988. "Comparative chemical composition of armed saltbush and four-wing saltbush." *J. Range Manage.* 41:401-403.

Guerrero, E.J. Undated. *Scientific, Standard, and Spanish Names of Woody Plants in South Texas* USDA Natural Resource Conservation Service: Rio Grande City. 1 page.

Hanselka, C.W., and J.M. Payne. 1989. *Landscaping with Native Plants to Promote Wildlife Habitat.* Texas Ag. Ext. Ser. Bull. L-2358. Texas A&M Univ.: College Station.

Hatch, S.L., and J. Pluhar. 1993. *Texas Range Plants.* Texas A&M Univ. Press: College Station. 326 pp.

Huston, J.E., B.S. Rector, L.B. Merrill, and B.S. Engdahl. 1981. *Nutritional Value of Range Plants in the Edwards Plateau Region of Texas.* Texas Ag. Exp. Stn. Bull. B-1357. Texas A&M Univ.: College Station. 16 pp.

Inglis, J.M. 1964. *A History of Vegetation on the Rio Grande Plain.* Bull. 45. Texas Parks and Wildlife Department: Austin. 122 pp.

Johnston, M.C. 1962. "Past and present grasslands of southern Texas and northern Mexico." *Ecology.* 44:456-466.

Kartesz, John T. 1994. *A Synonymized Checklist of the Vascular Flora of the United States, Canada, and Greenland.* 2nd ed. 2 vol. Timber Press: Portland, OR. 1,438 pp.

Lehouerou, H.N., and J. Norwine. 1988. "The ecoclimatology of south Texas." *In:* E.E. Whitehead, C.H. Hutchinson, B.N. Timmermann, and R.G. Varady, eds. *Arid Lands Today and Tomorrow*, pp. 417-443. Proc. Res. and Int. Dev. Symposium/Westview Press: Boulder, CO, and Bellhaven Press: London. 1,435 pp.

Lonard, R.I., J.H. Everitt, and F.W. Judd. 1991. *Woody plants of the Lower Rio Grande Valley, Texas.* Texas Memorial Museum. Univ. of Texas Press: Austin. 179 pp.

Loughmiller, C., and L. Campbell. 1984. *Texas Wildflowers.* Univ. of Texas Press: Austin. 271 pp.

Lynch, G.W. 1977. *Nutritive Value of Forage Species in the Rio Grande Plains of Texas for White-tailed Deer* (Odocoileus virginianus texanus) *and Domestic Livestock.* M.S. thesis. Texas A&M Univ: College Station. 83 pp.

Martin, A.C., H.S. Zim, and A.L. Nelson. 1951. *American Wildlife and Plants.* Dover Publications: New York.

McMahan, C.A., and J.M. Inglis. 1974. "Use of Rio Grande Plains brush types by white-tailed deer." *J. Range Manage.* 27(5):369–374.

Meyer, M.W., and R.D. Brown. 1985. "Seasonal trends in the chemical composition of Texas range plants in south Texas." *J. Range Manage.* 38(2):154–157.

Meyer, M.W., R.D. Brown, and M.W. Graham. 1984. "Protein and energy content of white-tailed deer diets in the Texas coastal bend." *J. Wildl. Manage.* 48(2):527–534.

Meyer, M.W., and W.H. Karasov. 1989. "Antiherbivore chemistry of *Larrea tridentata*, effects on woodrat (*Neotoma lepidum*), feeding and nutrition." *Ecology.* 70(4):953–961.

Miller, G.O. 1991. *Landscaping with Native Plants of Texas and the Southwest.* Voyageur Press: Stillwater, MN. 128 pp.

Montemayor, E., T.E. Fulbright, L. Brothers, B.J. Schat, and D. Cassels. 1991. "Long term effects of rangeland disking on white-tailed deer browse in south Texas." *J. Range Manage.* 44:246–248.

Natural Heritage Policy Research Project. 1978. *Preserving Texas Natural Heritage.* Report No. 31. Lyndon B. Johnson School of Public Affairs. Univ. of Texas: Austin. 34 pp.

Natural Resources Conservation Service. 1981. *Webb County Area Forage Analysis.* Unpublished data.

Natural Resources Conservation Service. 1989. *Pecos County Area Forage Analysis.* Unpublished data.

Natural Resources Conservation Service. 1990. *San Angelo Area Forage Analysis.* Unpublished data.

Nelle, S. 1984. "Key food plants for deer–South Texas." *In: Proceedings of the 1984 International Rancher Roundup*, pp. 281–291. Texas Ag. Ext. Serv. Texas A&M Univ.: College Station.

Nokes, J. 1986. *How to Grow Native Plants of Texas and the Southwest.* Gulf Publishing Company: Houston. 404 pp.

Quinton, D.A., R.G. Horejsi, and J.T. Flinders. 1979. "Influence of brush control on white-tailed deer diets in north-central Texas." *J. Range Manage.* 32(2):93–97.

Richardson, A. 1995. *Plants of the Rio Grande Delta.* Univ. of Texas Press: Austin. 332 pp.

Richardson, C.L. 1990. *Factors Affecting Deer Diets and Nutrition.* Texas A&M Univ.: College Station. 6 pp.

Scrifes, C.J. 1980. *Brush Management: Principles and Practices for Texas and the Southwest*. Texas A&M Univ. Press: College Station. 360 pp.

Simpson, B.J. 1988. *A Field Guide to Texas Trees*. Texas Monthly Press: Austin. 372 pp.

Spalinger, D.E., D.J. Morting, L.A. Newton, and C.K. Pape. 1991. "Foraging Ecology of Small Ruminants in South Texas." *In: Risks in Ranching*, pp. 26–30. Texas Ag. Exp. Stn. Bull. CPR 4870-4878. Texas A&M Univ: College Station.

Steuter, A.A., and H.A. Wright. 1980. "White-tailed deer densities and brush cover on the Rio Grande Plains." *J. Range Manage.* 33(5):328–331.

Stubbendieck, J., S.L. Hatch, and K.J. Hirsch. 1986. *North American Range Plants*. Univ. of Nebraska Press: Lincoln. 465 pp.

Texas Forest Service. 1963. *Forest Trees of Texas: How to Know Them*. Bull. 20. Texas A&M Univ.: College Station. 156 pp.

Texas Parks and Wildlife Department. 1994. *Forage Analysis of Selected Shrubs on the Chaparral Wildlife Management Area*. Unpublished data.

Texas Parks and Wildlife Department. 1995. *Forage Analysis of Selected Shrubs on the Chaparral Wildlife Management Area*. Unpublished data.

Varner, L.W., L.H. Blankenship, and G.W. Lynch. 1977. "Seasonal changes in nutrient value of deer food plants in south Texas." *Proc. Ann. Conf. Southeast. Assoc. Fish and Wildlife Agencies.* 31:99–106.

Vines, R.A. 1960. *Trees, Shrubs, and Woody Plants of the Southwest*. Univ. of Texas Press: Austin. 1,104 pp.

Wasowski, S., and A. Wasowski. 1991. *Native Texas Plants: Landscaping Region by Region*. Gulf Publishing Company: Houston. 406 pp.

Weniger, D. 1988. *Cacti of Texas and Neighboring States*. Univ. of Texas Press: Austin. 356 pp.

White, L.D. 1979. *Forage Analysis of Selected Plants on the Willingham Ranch in Uvalde County*. Unpublished data.

Whitehouse, E. 1962. *Common Fall Flowers of the Coastal Bend of Texas*. Rob & Bessie Welder Wildlife Foundation: Sinton, TX. 116 pp.

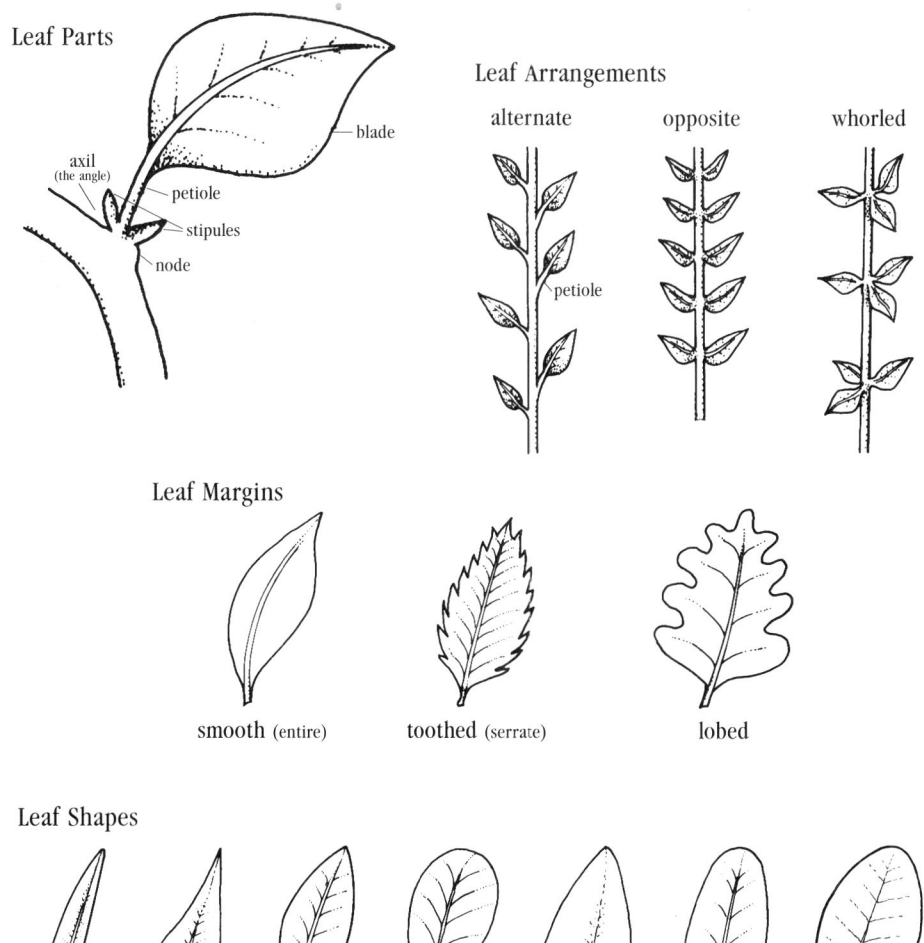

Leaf Parts
- blade
- axil (the angle)
- petiole
- stipules
- node

Leaf Arrangements
- alternate
- opposite
- whorled
- petiole

Leaf Margins
- smooth (entire)
- toothed (serrate)
- lobed

Leaf Shapes
- linear
- lanceolate
- elliptic
- spatulate
- ovate
- oblong
- oval

Drawings from *Texas Wildflowers: A Field Guide*
by Campbell and Lynn Loughmiller

Leaf Types

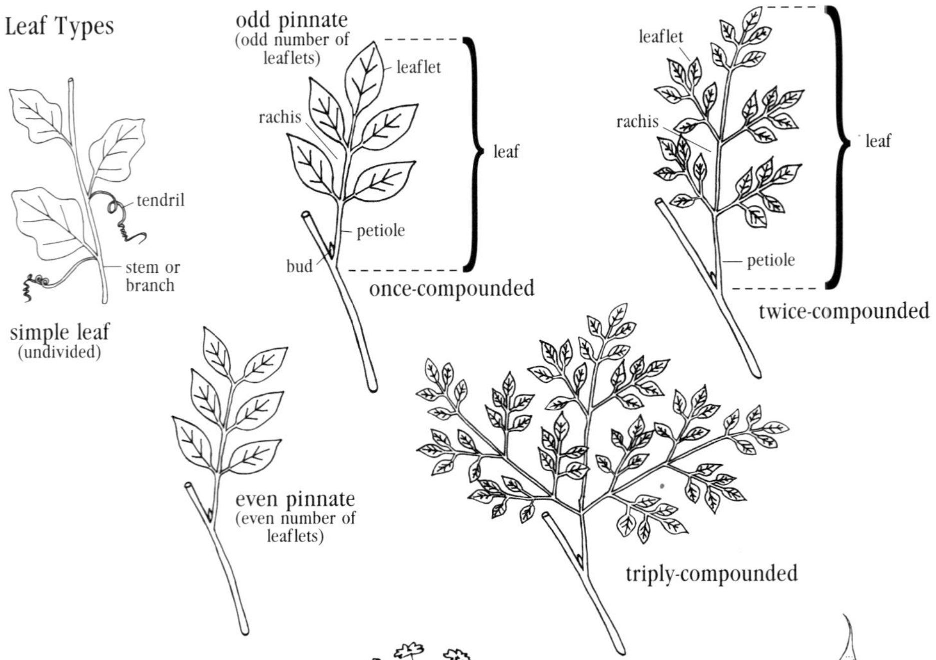

odd pinnate
(odd number of leaflets)

leaflet

rachis

leaf

petiole

bud

once-compounded

leaflet

rachis

leaf

petiole

twice-compounded

tendril

stem or branch

simple leaf
(undivided)

even pinnate
(even number of leaflets)

triply-compounded

Drawings of Leaf Types, Flower Stalk and Legume from *Wildflowers of the Big Thicket, east Texas, and western Louisiana* by Geyata Ajilvsgi. Illustrations by Martha Bell

Legume

bean

joint

pedicel

peduncle

Flower Stalk

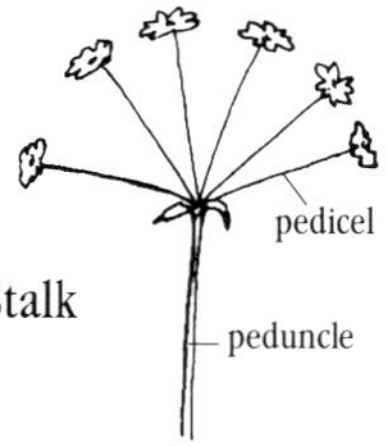

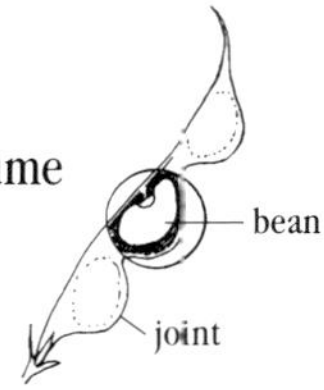

Flower Parts

petal

pistil
- stigma
- style
- ovary

anther
filament
stamen

sepal

Flower Types

spike

raceme

catkin

Drawings of Flower Parts and Flower Types from *Texas Trees: A Friendly Guide* by Paul W. Cox and Patty Leslie, illustrations by Gloria Merlo and Sara Harrison

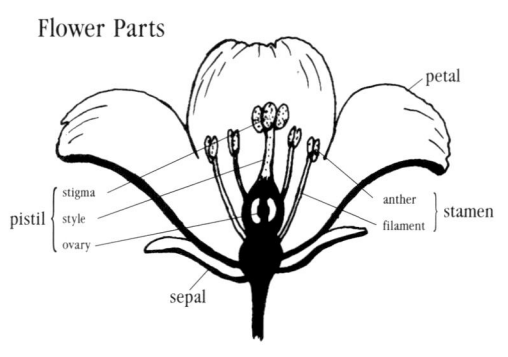
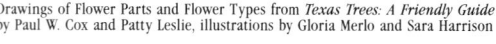
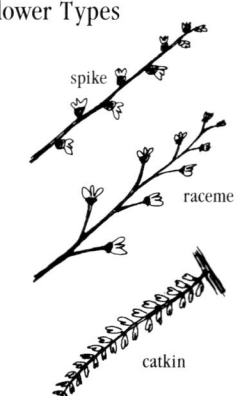

102

lime elm · 80
lime prickly-ash · 32
littleleaf sumac · 46, 95
live oak · 68, 95
lluvia de oro · 20
lotebush · 30, 97
Lycium berlandieri · 38

Mahonia trifoliata · 48
manzanillo · 50
manzanita · 50
mejorana · 70, 84
mescalbean · 64
mesquite · 14, 94
Mexican olive · 52
Mexican persimmon · 58
Mint Family · 70
monte cristo · 84
Mormon tea · 60
mountain laurel · 64, 94
myrtle croton · 62

narrowleaf forestiera · 72, 95
Nightshade Family · 38
nopal · 2

OLEACEAE · 72
Olive Family · 72
olmo · 80
Opuntia engelmannii · 2
Opuntia leptocaulis · 4
oreja de raton · 62

paisano bush · 48
palma loca · 44
palma pita · 44
palo amarillo (2) · 48, 86
palo blanco (2) · 40, 82
palo verde · 18, 97
panalero (2) · 54, 72
Parkinsonia aculeata · 20
Parkinsonia texana · 18

pencil cactus · 4
Persimmon Family · 58
pita · 44
Pithecellobium ebano · 22
popote · 60
popotillo · 60
prickly pear · 2
Prosopis glandulosa · 14
purple sage · 78

Quassia Family · 36
Quercus virginiana · 68

rat-tail cactus · 4
retama · 20, 97
RHAMNACEAE · 26, 28, 30, 74, 76
Rhus microphylla · 46
rockbush · 66
RUTACEAE · 32

saffron-plum bumelia · 34
Salvia ballotiflora · 70
Sapodilla Family · 34
SAPOTACEAE · 34
Schaefferia cuneifolia · 54
Schaffner acacia · 24
SCROPHULARIACEAE · 78
scrub elm · 80
scrub sumac · 46
shad-scale · 56
shrubby blue sage · 70, 95
shrubby blue salvia · 70
Sideroxylon celastrinum · 34
SIMAROUBACEAE · 36
small-leaved sumac · 46
soapbush · 90
SOLANACEAE · 38
Sophora secundiflora · 64
southern rock elm · 80
southwest bernardia · 62, 93
Spanish bayonet · 44
Spanish dagger · 44

Notes

Notes

About the Authors

Richard B. (Rick) Taylor, Jimmy Rutledge and Joe G. Herrera were born and raised in south Texas. They are wildlife biologists for the South Texas Regulatory District, Wildlife Division, Texas Parks and Wildlife Department. Taylor and Rutledge graduated with Bachelor of Science degrees in Range and Wildlife Management from Texas A&I University (Texas A&M–Kingsville). Taylor's primary responsibility includes monitoring wildlife populations and assisting landowners throughout upper south Texas. Rutledge is the technical guidance biologist for the south Texas area. Herrera, an avid photographer, graduated from Texas A&M University with a Bachelor of Science degree in Wildlife and Fisheries Science and is currently district supervisor for the South Texas Regulatory District.